Acidizing Fundamentals

Bert B. Williams

Manager, Drilling Technology

Esso Exploration, Inc.

John L. Gidley

Technical Advisor

Exxon Company, U.S.A.

Robert S. Schechter

Professor, Dept. of Petroleum Engineering

University of Texas

Henry L. Doherty Memorial Fund of AIME

Society of Petroleum Engineers of AIME

New York   1979   Dallas

ISBN 0-89520-205-0

Contents

1. History of Acidizing — 1

  1.1 Introduction — 1

  1.2 Frasch Acidizing Patent — 1

  1.3 Early Scale-Removal Treatments — 2

  1.4 Beginning of the Modern Era of Acidizing — 2

  1.5 Formation of Acidizing Companies — 2

  1.6 Early Sandstone Acidizing History — 2

2. Acidizing Methods — 5

  2.1 Introduction — 5

  2.2 Description of Acid Treatments — 5

  2.3 Theoretical Productivity Improvement From Acidization — 5

3. Acid Types and the Chemistry of Their Reactions — 10

  3.1 Introduction — 10

  3.2 Acid Systems and Considerations in Their Selection — 10

  3.3 Stoichiometry of Acid Carbonate Reactions — 12

  3.4 Equilibrium in Acid-Carbonate Reactions — 14

  3.5 Stoichiometry of Acid-Sandstone Reactions — 16

  3.6 Equilibrium in Acid-Sandstone Reactions — 17

4. Reaction Kinetics — 19

  4.1 Introduction — 19

  4.2 Surface Reaction Kinetics — 19

  4.3 Mass Transfer in Acid Solutions — 21

  4.4 Models for Heterogeneous Reactions in Laminar Flow Systems — 23

5. Acid Fracturing Fundamentals — 29

  5.1 Introduction — 29

  5.2 Fracture Geometry — 29

  5.3 Acid Penetration Distance — 30

  5.4 Fracture Conductivity — 35

6. Acid Fracturing Treatment Models — 38

  6.1 Introduction — 38

  6.2 Dynamic Fracture Geometry — 38

  6.3 Equations Used to Define Fracture Geometry — 40

  6.4 Fluid Temperature in the Fracture — 42

  6.5 Acid Penetration Along the Fracture — 44

  6.6 Fracture Conductivity — 49

7. Acid Fracturing Treatment Design — 53

  7.1 Introduction — 53

  7.2 Determination of Formation Matrix and Fluid Properties — 53

  7.3 Selection of Variable Design Parameters — 58

  7.4 Calculation of Fracture Geometry and Acid Penetration Distance — 60

  7.5 Prediction of Fracture Conductivity and Stimulation Ratio — 62

  7.6 Select the Most Economic Treatment — 65

  7.7 Design Hints for Unconventional Treatments — 65

  7.8 Design of Acid Fracturing Treatments To Remove Near-Wellbore Damage — 66

8. Models for Matrix Acidizing — 68

  8.1 Introduction — 68

  8.2 Description of a Model for Matrix Acidization — 68

  8.3 Application of the Matrix Acidizing Model to Retarded Acid Systems — 69

  8.4 Application of the Model to Slowly Reacting Systems — 70

  8.5 Application of the Model to Fast-Reacting Systems — 70

  8.6 Prediction of Acid Penetration Into Sandstones — 71

  8.7 Model of Wormhole Growth in Carbonate Acidization — 73

9. Matrix Acidizing of Sandstones — 76

  9.1 Introduction — 76

  9.2 Description of a Sandstone Acidizing Treatment — 76

  9.3 Mechanism of Acid Attack — 76

  9.4 Prediction of Radius of Acid Reaction — 78

  9.5 Productivity After Sandstone Acidizing — 79

  9.6 Design Procedure for Sandstone Acidizing Treatments — 81

  9.7 Common Mistakes in Application of Acid Treatments — 83

  9.8 Future Trends in Sandstone Acidizing — 83

10. Matrix Acidizing of Carbonates — 86

  10.1 Introduction — 86

10.2	Description of a Matrix Acid Treatment in a Carbonate	86
10.3	Mechanism of Acid Attack	86
10.4	Prediction of Radius of Reaction	87
10.5	Acids Used in Matrix Treatments	88
10.6	Design Procedure for Matrix Acidization of Carbonates	88
10.7	Novel Matrix Acid Treatments for Carbonates	89

11. Acid Additives — 92

11.1	Introduction	92
11.2	Corrosion Inhibitors	92
11.3	Surfactants	95
11.4	Mutual Solvents	96
11.5	Friction Reducers	97
11.6	Acid Fluid-Loss Additives	97
11.7	Diverting Agents	99
11.8	Complexing Agents	100
11.9	Cleanup Additives	102

12. Acidizing Economics — 104

12.1	Introduction	104
12.2	Cash Flow Analysis	104
12.3	Payout Period	104
12.4	Profit-To-Investment Ratio	105
12.5	Discounted Cash Flow Techniques	106
12.6	Use of Economic Yardsticks	106

Appendix: Conversion Factors	**107**
Nomenclature	**110**
Bibliography	**112**
Author-Subject Index	**118**

SPE Monograph Series

The Monograph Series of the Society of Petroleum Engineers of AIME was established in 1965 by action of the SPE Board of Directors. The Series is intended to provide members with an authoritative, up-to-date treatment of the fundamental principles and state of the art in selected fields of technology. The work is directed by the Society's Monograph Committee, one of 40 national committees, through a Committee member or members designated as Monograph Coordinator. Technical evaluation is provided by the Monograph Review Committee. Below is a listing of those who have been most closely involved with the preparation of this book.

Monograph Coordinators

Roscoe C. Clark, Continental Oil Co., Houston

W. C. Hardy, Sun Co., Dallas

Michael Prats, Shell Development Co., Houston

Monograph Review Committee

Roscoe C. Clark, chairman, Continental Oil Co., Houston

W. T. Strickland, Shell Development Co., Houston

C. F. Smith, Dowell, Tulsa

R. M. Lasater, Halliburton Services, Duncan

Michael Prats, Shell Development Co., Houston

SPE Monograph Staff

Jim McInnis	Ann Gibson	Sondra Stewart
Publications Manager/Editor	Production Manager	Project Editor

Acknowledgments

We want to thank Exxon Co., U.S.A., and Exxon Production Research Co. for allowing us to use company time and support facilities in the preparation of this monograph. Also, we appreciate the help of various SPE-AIME members and personnel, particularly the Monograph Review Committee, in providing helpful suggestions on organization and editing of the monograph.

Many people have contributed to the successful completion of this monograph; therefore, it is impossible to recognize them all. Both D. E. Nierode and E. J. Novotny of Exxon Production Research Co. assisted in updating technology during our long writing and review period. During preparation, the manuscript was used as a text for a graduate course at The U. of Texas at Austin. The students taking that course provided many helpful comments and verified example calculations. They are Essam M. Abdallah, Hosny H. Abdel Kareem, Amr Mohammed Badawy, Paul Michael Bommer, Tommy James Drescher, Doan Phu Duy, Rufus Oladipo Elemo, Nguyen Thi Hang, In Sul Hyun, Kazem Javanmardi, Jeoung Soo Kim, Ser Yuen Kwong, Wei-Quo Liu, Motoyoshi Naiki, Mehmet Melih Oskay, Adolfo Antonio Rosales, Subrata Sen, and Eugenia Vasquez. Finally, this monograph could not have been prepared without the able assistance of the Exxon Production Research Co. editorial, word processing, and illustration departments. In particular, the efforts of N. H. Parker and P. J. Henry were essential.

Foreword

In recent years, stimulation techniques have become increasingly complex and a better understanding of the individual mechanisms that contribute to the over-all treatment effectiveness has evolved. Much of this development has been at the research level and is not totally assimilated into general field practice.

This monograph has been prepared to serve as a basic reference for both the field and research engineer interested in developing a fundamental understanding of the acidizing process and procedures for designing acid treatments to increase well productivity. Accordingly, we have attempted to structure the monograph to serve both groups. Some chapters provide a comprehensive review of the science and technology that serve as bases for understanding the fundamentals of acid stimulation. Others are written for the field engineer and stress design procedures and the proper selection of fluids and additives. These procedures employ practices in wide-spread use in the United States and, to a lesser extent, in use throughout the free world.

Topics that were covered in detail by G. C. Howard and C. R. Fast in *Hydraulic Fracturing,* Volume 2 of the SPE-AIME Monograph Series, are not repeated. In particular, the reader interested in field application of acid fracturing techniques should read Chapters 1, 2, 3, and 8 in *Hydraulic Fracturing*. These chapters are respectively titled Introduction, Theories of Hydraulic Fracturing, Determination of Wells Applicable for Fracturing, and Mechanical Equipment for Hydraulic Fracturing. These chapters lay the groundwork for material on acid fracturing presentéd in this monograph.

We strongly urge all readers of the *Acidizing Fundamentals* monograph to start with Chapters 2 and 3, which review acidizing methods and the chemistry of acid reactions. The reader can then proceed to Chapters 5 and 7 for a discussion of acid fracturing fundamentals and design procedures and to Chapter 9 or 10 for a discussion of the design of matrix acid treatments for sandstone and carbonate formations. Chapters 4, 6, and 8 are fundamental chapters that will be of primary interest to the reader requiring more detail. Although Chapters 4, 6, and 8 present the groundwork for design procedures given elsewhere, one should not have to read these chapters before reading the design sections. Chapters 1, 11, and 12 cover specialized topics that will be of general interest (history of acidization, acid additives, and economics).

Wherever possible, examples have been provided to clarify the application of the models or concepts presented in this monograph. In some instances, models used in the examples are not the most comprehensive available since we have limited our use to those that require only simple arithmetic manipulations with a hand calculator. Throughout the monograph, equations are given in nondimensional form unless otherwise noted. Because of the familiarity of our audience with conventional engineering units, they are used in all examples.

Chapter 1

History of Acidizing

1.1 Introduction

However one evaluates the history of well stimulation processes, acidizing must be considered among the oldest techniques still in modern use. Only nitro-shooting predates it. Other techniques such as hydraulic fracturing were developed much more recently. Knowledge of acidizing as a well stimulation method began in the last century.

Earliest records indicate that the first acid treatments were probably performed in 1895.[1]* Most interestingly, hydrochloric acid, the agent commonly preferred today, was employed in these tests. Herman Frasch, who at the time was chief chemist at Standard Oil Co.'s Solar Refinery at Lima, Ohio, is credited with having invented the technique.[2] Of all the patents on acidizing, the first — the one issued to Frasch on March 17, 1896,[3] — is perhaps the most instructive. Recorded in that brief document are many of the elements of present-day acid treatments. Not as important technically, but of historical interest, is a similar patent employing sulfuric acid, obtained by John W. Van Dyke, general manager of Solar Refinery, and a close friend of Frasch. Subsequently, Frasch and Van Dyke each assigned the other one-half interest in their respective patents, possibly revealing some doubt in the inventors' minds as to which process would be successful.[4]

1.2 Frasch Acidizing Patent

The Frasch patent involved a reagent (hydrochloric acid) that would react with limestone to produce soluble products — carbon dioxide and calcium chloride — that then could be removed from the formation as the well fluids were produced. In contrast, the Van Dyke process, using sulfuric acid, produces insoluble calcium sulfate, which is capable of plugging the formation.

Frasch's concept of the acidizing process contained many of the elements of current techniques. In his patent, he described "a new and superior method based upon chemical action . . . (in which) a chemical reagent . . . attacks the limestone-rock." Frasch anticipated the need "to put (the acid) under strong pressure" so that it might be "pressed into the rock and made to act upon the same at a distance

from the original well-hole." He also stated that "long channels can be formed" in the process. The patent called for "the use of commercial muriatic or hydrochloric acid (which contains from thirty to forty percent by weight of the acid gas HCl)." The use of an afterflush was anticipated in the following statement: "It is advantageous to displace it (the acid) and cause it to penetrate further into the rock by forcing a neutral or cheap liquid, such as water, into the well."

Frasch was unsure* how completely the acid would be spent. To avoid returning unspent acid to the wellbore, he proposed "to introduce an alkaline liquid (preferably milk of lime)" to neutralize any remaining trace of acid. The advantage of neutralization, he went on, "is to avoid the danger of corroding the subsequently used apparatus." Realizing that his process required the injection of a highly corrosive material, Frasch proposed the use of pipe that was "enameled or lead lined" or "otherwise made proof against corrosion." He also corresponded with a rubber company to determine if lining the pipe with rubber would be practical. Finally, Frasch anticipated the need for a rubber packer to pack off the annulus and force the acid into the formation to be treated.

Frasch and Van Dyke conceived the acidizing process as a means to increase production in oil wells in the Lima, Ohio, area, which at the time accounted for almost one-third of domestic production.[5] Some of the wells in this area produced little in comparison with more prolific offset wells. The inventors reasoned that production could be increased in these less productive wells if more conductive paths to the wellbore could be created by acid.[6] Evidently, the method was applied vigorously and with success. The *Oil City Derrick* carried a feature story[1] that explained the process in detail. The first application was described as follows: "Two months ago, a practical test of this process was made on a well on the Crosley Farm owned by the Ohio Oil Company near Lima, Ohio. There were sixty-five barrels of acid used in the well on the Crosley Farm. The channels in the oil rock in this well were so tight that even with a pressure of between

---

*Secondary references[2] often mention the year 1894, but that date appears to be unsubstantiated.

*It appears his patent application was filed approximately 2 months before the first field test.

800 and 900 pounds per square inch the rock would take barely a barrel of acid and water per hour. However, as the acid began to perform its work, the channels in the rock were gradually increased so that at the finish the rock would readily take six barrels per hour.''

''Since doctored, this well has been pumped some forty days. The oil was increased 300% and the gas over 400%. The increase has been permanent as the well is holding up.'' The article ended with the statement that ''While this process may seem very simple, yet it has required considerable ingenuity and skill and a practical knowledge of oil wells as well as chemical facts to develop the thought and bring it to its present perfection so that now an oil well can as readily be doctored with acid as it can be torpedoed, and results obtained that cannot be hoped for from the use of nitroglycerin.''[1]

Although the new process was used many times within the next year or two, for some reason, not entirely clear from the historical record, its use declined and no evidence of acidizing is available during the ensuing 30 years.

1.3 Early Scale-Removal Treatments

The next significant use of hydrochloric acid in well treatment took place in 1928 in Oklahoma by the Gypsy Oil Co., a subsidiary of Gulf Oil Co.[7] The problem facing Gypsy was a calcareous scale deposited in the pipe and equipment in certain wells producing from a sandstone formation. For a technique useful for scale removal, Gypsy sought the advice of the Mellon Institute. Dr. Blain Wescott reported on behalf of the Institute suggesting the use of hydrochloric acid as a solvent for the scale. Interestingly, Wescott's report recommended the use of an inhibitor, Rodine No. 2, which at that time was being employed in steel mills.[8,9] Apparently, a patent application was not filed on the use of the inhibitor since it was thought to be old art merely adapted from the steel industry.*

Gypsy's use of hydrochloric acid to dissolve scale was successful. Its use declined, however, in the early 1930's with the decline in oil prices.[7]

1.4 Beginning of the Modern Era of Acidizing

What might be described as the modern era of acidizing began in 1932 in discussions between the Pure Oil Co. and the Dow Chemical Co. Pure had oil property in Michigan and an active exploration program in the area. Dow had brine wells in the same area. Pure requested operational information from Dow to support its program. Dow, having no interest in oil production at that time, agreed to make its brine-well files available to Pure. In subsequent discussions between Pure and Dow, Pure's geologist, W. A. Thomas, knowing that hydrochloric acid would react with limestone, and apparently unaware of the earlier work by Frasch, suggested that well productivity from limestone formations might be improved by acid treatment. John Grebe, who was in charge of Dow's Physical Research Laboratory, concurred and mentioned Dow's experiences in treating brine

wells (completed in sandstone) with acid. Finally, Pure proposed acidizing one of its own wells. The test site was decided upon and on Feb. 11, 1932, the Fox No. 6 well in Section 13, Chippewa Township, Isabella County, Mich., was treated with 500 gal of hydrochloric acid. Acid was brought to the wellsite on a tank wagon equipped with a wooden tank 36 in. in diameter and 12 ft long. To this acid, 2 gal of an arsenic acid inhibitor were added, at the suggestion of Grebe, to reduce corrosion of the tubing. The acid was transferred from the tank truck to the wellbore by siphoning with a garden hose. About half of the 500 gal of acid was siphoned into the tubing. This was followed by 6 bbl of oil pumped into the tubing (with a hand-operated pump) after the acid. The well was shut in over night and swabbed in the next morning. A large quantity of emulsion was removed. The remaining acid was siphoned into the tubing and displaced by oil flush.[10]

This appears to be the first use of inhibited acid in a limestone formation. The well, which was dead before treatment, subsequently produced as much as 16 B/D. Other wells were later treated with acid. Some responded better than the first.[11-13]

1.5 Formation of Acidizing Companies

Interest in acidizing spread rapidly and companies were formed to provide this service.[14-19] Dow originally handled well treatments through its Well Service Group — whose main responsibility was Dow's brine wells. The oil well service activities of this group grew rapidly and it became necessary for Dow to form a new subsidiary on Nov. 19, 1932, to handle this business. Taking its name from the Dow Well Service Group, the first two words were combined to Dowell, but the pronunciation remained the same; thus, Dowell Inc. was formed.[10,20]

Other companies originating during this period include the Oil Maker's Co.,* organized in Michigan in June 1932,[9] and the Chemical Process Co.,** begun in Texas in Oct. 1932.[21] Williams Brothers Treating Corp.* was formed in Oklahoma in April 1932. Each of these companies did a rapidly expanding business. In March 1935, Halliburton Oil Well Cementing Co. began acidizing oil wells commercially.[8]

1.6 Early Sandstone Acidizing History

Success with acids in limestone brought many to think of treatments useful for sandstone formations. On March 16, 1933, Jesse Russell Wilson, with the Standard Oil Co. of Indiana, filed a patent application on a process for treating sandstone formations with hydrofluoric acid.[22]† In this process, hydrofluoric acid was generated either in the wellbore or in the formation to avoid the danger of handling it at the surface. Wilson recognized the ability of hydrofluoric acid

*No longer in existence.

**The company operated under this name until 1958, when it became a part of Byron-Jackson, Inc.

†A similar patent (U.S. Patent No. 2,094,479) was filed on the same date by James G. Vandergrift but was issued much later (Sept. 28, 1937) on a process employing a mixture of mineral acid and hydrofluoric acid.

*Scientific publications describing the use of inhibitors in acid pickling in the steel industry date from 1845.[9]

to react with sandstone and silicious materials. He must have envisioned his treatment to be especially useful for formation damage removal because he described the problem to be treated as follows: "Finely divided sand, other silicious and miscellaneous debris tend to be deposited by the fluid flowing toward the base of the well, thereby clogging up the pores or passages in the geological formation immediately surrounding the base of the well with the result that resistance to flow is greatly increased. It has occurred to me that one method of rectifying this situation is to dissolve out this deposited material by the use of a suitable reagent. In the case of sand, one suitable reagent is hydrofluoric acid or hydrogen fluoride which reacts with the sand . . . producing water and silicon tetrafluoride, the latter being a gas. The principal difficulty with this procedure is that hydrofluoric acid is an extremely dangerous material to handle. The risk encountered in introducing it into an oil well would be so great that I do not believe it has ever actually been attempted."[22]

However, others were already planning to attempt it. In early 1933, A. M. McPherson from Wichita Falls, Tex., apparently unaware of Wilson's work, approached Halliburton's management about the use of hydrofluoric acid in sandstone. He believed that it would be as effective as hydrochloric acid had been in limestone. McPherson was hired by Halliburton to work with its chemistry laboratory to make hydrofluoric acid applicable to oil well use. Several months later, a well belonging to the King Royalty Co. was selected for the test. It was Wilson B-24, Block 88, American Tribune New Colony Subdivision in Archer County, Tex., 5 miles north and 2 miles west of Archer City. It was 1,532 ft deep and had 11 ft of open-hole producing zone.[23]

On May 3, 1933, less than 2 months after Wilson's application had reached the U.S. Patent Office, a mixture of hydrofluoric and hydrochloric acid was pumped into the well. Special precautions were taken for handling the acids — rubber suits, rubber gloves, and respirators were provided for the personnel on the job. In addition to McPherson, Halliburton also had Phil Montgomery and Hayden Roberts on the test. These men spent 3 weeks following up the job after the treatment.

No records are at hand to determine the concentration of hydrofluoric and hydrochloric acids used in this first treatment. Nonetheless, the results were disappointing. The acids dissolved the calcareous matrix of the sandstone and left in the wellbore a large quantity of unconsolidated sand that later was removed by swabbing and pumping. According to these observers, "The reaction products of the acids on the sand seemed to have a plugging effect on the permeability of the formation." Halliburton subsequently discontinued work along this line and did not offer hydrofluoric-hydrochloric acid mixtures commercially until the middle 1950's.

Although Wilson's patent was silent on the concentration of hydrofluoric acid to be used for best results, it recognized many of the problems that attend this treatment. Wilson recommended the use of excess hydrochloric acid to avoid reaction of silicon tetrafluoride with water "to form insoluble or gelatinous, silicic, and hydrofluosilicic acids which

clog up the formation." He also suggested the use of an acid preflush to prevent deposition of insoluble or gelatinous materials. In his example for the in-situ generation of hydrofluoric acid, Wilson used sodium fluoride reacted with hydrochloric acid. It was not until some 30 years later that the damaging nature of the sodium salt was fully appreciated.[24] Aside from this, the Wilson patent is an amazingly comprehensive statement of the problem, and it taught techniques still in use today.

The first commercial use of mixtures of hydrochloric and hydrofluoric acid was begun in 1940 under the auspices of Dowell.[25-29] This product, called Mud Acid, was developed to dissolve the drilling mud deposited as a filter cake during the rotary drilling process. First treatments occurred in the Gulf Coast area and were sufficiently successful to warrant widespread interest and expanded use. The treatment, subsequently modified and improved, has continued to be used.

References

1. "A Great Discovery," *Oil City Derrick* (Oct. 10, 1895).

2. Putnam, S. W.: "Development of Acid Treatment of Oil Wells Involves Careful Study of Problems of Each," *Oil and Gas J.* (Feb. 23, 1933) 8.

3. Frasch, H.: "Increasing the Flow of Oil Wells," U.S. Patent No. 556,669 (March 17, 1896).

4. Fitzgerald, P. E.: "A Review of the Chemical Treatment of Wells," *J. Pet. Tech.* (Sept. 1953) 11-13.

5. Hidy, R. W. and Hidy, M. E.: *Pioneering in Big Business-History of Standard Oil Company (New Jersey) 1882-1911*, Harper & Brothers, New York (1955) 156.

6. McCann, B. E.: "Chemistry Dons a Hard Hat . . . The Saga of Acidizing Service," *Drilling* (July 1968) 35.

7. Chapman, M. E.: "Some of the Theoretical and Practical Aspects of the Acid Treatment of Limestone Wells," *Oil and Gas J.* (Oct. 12, 1933) 10.

8. Hathorn, D. H.: personal communication, Halliburton Services, Duncan, Okla. (June 7, 1971).

9. "The Dow Chemical Company v. Halliburton Oil Well Cementing Company," Opinion of Circuit Court of Appeals, Sixth Circuit, U.S. Patent Quarterly 90.

10. Fitzgerald, P. E.: "For Lack of a Whale," (unpublished notes on a history of Dowell) Dowell, Houston.

11. "Acid Treatment Becomes Big Factor in Production," *Oil Weekly* (Oct. 10, 1932) 57.

12. Newcombe, R. B.: "Acid Treatment for Increasing Oil Production," *Oil Weekly* (May 29, 1933) 19.

13. Putnam, S.: "The Dowell Process to Increase Oil Production," *Ind. Eng. Chem.* (Feb. 20, 1933) 51.

14. Putnam, S. W. and Fry, W. A.: "Chemically Controlled Acidation of Oil Wells," *Ind. Eng. Chem.* (1934) **26,** 921.

15. "Chemical Treatment Halts Junking Breckenridge Wells," *Oil Weekly* (Feb. 13, 1932) 40.

16. "North Louisiana Operators Pleased with Acid Results," *Oil Weekly* (Dec. 19, 1932) 70.

17. Clason, C. E. and Staudt, J. G.: "Limestone Reservoir Rocks of Kansas React Favorably to Acid Treatment," *Oil and Gas J.* (April 25, 1935) 53.

18. Bancroft, D. H.: "Acid Tests Increase Production of Zwolle Wells," *Oil and Gas J.* (Dec. 22, 1932) 42.

19. Moore, W. W.: "Acid Treatment Proved Beneficial to North Louisiana Gas Wells," *Oil Weekly* (Oct. 29, 1934) 31.

20. "Chemical Company Forms Company to Treat Wells," *Oil Weekly* (Nov. 28, 1932) 59.

21. Pitzer, P. W. and West, C. K.: "Acid Treatment of Lime Wells Explained and Methods Described" *Oil and Gas J.* (Nov. 22, 1934) 38.

22. Wilson, J. R.: "Well Treatment," U.S. Patent No. 1,990,969 (Feb. 12, 1935).

23. Roberts, H.: "Creative Chemistry — A History of Halliburton Laboratories 1930-1958," Halliburton Oil Well Cementing Co., Duncan, Okla. (Jan. 7, 1959).

24. Smith, C. F. and Hendrickson, A. R.: "Hydrofluoric Acid Stimulation of Sandstone Reservoirs," *J. Pet. Tech.* (Feb. 1965) 215-222; *Trans.,* AIME, **234.**

25. "Mud Acid," *The Acidizer,* Dowell Inc. (June 1940).

26. Morian, S. C.: "Removal of Drilling Mud from Formation by Use of Acid" *Pet. Eng.* (May 1940) 117.

27. Cannon, G. E.: "Mud Acid and Formation Washing Agents," Topic No. 31-B, Standard Oil Co. (New Jersey) special reports (1942).

28. Herrington, C. G.: "Recent Developments in the Chemical Treatment of Wells," Topic No. 31-C Standard Oil Co. (New Jersey) special reports (1942).

29. Flood, H. L.: "Current Developments in the Use of Acids and Other Chemicals in Oil Production Problems," *Pet. Eng.* (Oct. 1940) 46.

Chapter 2

Acidizing Methods

2.1 Introduction

Acids derive their utility in well stimulation from their ability to dissolve formation minerals and foreign material, such as drilling mud, that may be introduced into the formation during well drilling or workover procedures. The extent to which the dissolution of these materials will increase well productivity depends on a number of factors, including the acidizing method chosen. The normally used acidizing techniques fall broadly into three categories: acid washing, matrix acidizing, and acid fracturing.

2.2 Description of Acid Treatments

Acid washing is an operation designed to remove acid-soluble scales present in the wellbore or to open perforations. It may involve nothing more than spotting a small quantity of acid at the desired position in the wellbore and allowing it to react, without external agitaton, with scale deposits or the formation. Alternatively, the acid may be circulated back and forth across the perforations or formation face. Circulation may accelerate the dissolution process by increasing the transfer rate of unspent acid to the wellbore surface.

Matrix acidizing is defined as the injection of acid into the formation porosity (intergrannular, vugular, or fracture) at a pressure below the pressure at which a fracture can be opened. The goal of a matrix acidizing treatment is to achieve, more or less, radial acid penetration into the formation. Stimulation is usually accomplished by removing the effect of a formation permeability reduction near the wellbore (damage) by enlarging pore spaces and dissolving particles plugging these spaces. Matrix acidizing is often most useful where acid fracturing cannot be risked because a shale break or other natural flow boundaries must be maintained to minimize or prevent water or gas production. When performed successfully, matrix acidizing often will increase oil production without increasing the percentage of either water or gas produced.

Acid fracturing is the injection of acid into the formation at a pressure high enough to fracture the formation or open existing fractures. Stimulation is achieved when a highly conductive flow channel remains open after the treatment. This channel is formed by acid reaction on the acid-soluble walls of the fracture. A high fracture conductivity can exist after the treatment if the etched fracture faces do not seal together when pressure is released and the fracture closes. The length of the conductive fracture created in acid fracturing is determined by a combination of the rate of acid reaction and the rate of fluid loss from the fracture to the formation. This conductive-fracture length is the factor limiting stimulation.

In addition to the uses described, acids are sometimes used for the following purposes:

- As a spearhead when fracturing to dissolve fine particles formed in the perforating process, thereby allowing the fracturing fluid to enter all perforations.
- To break emulsions in the formation when the emulsion is sensitive to a pH reduction or is stabilized by fine particles that the acid can dissolve.
- To break an acid-sensitive viscous gel used in a fracturing treatment if it does not break after the treatment.
- As a preflush before squeeze cementing.

Techniques for designing treatments for these four possible uses of acid are poorly defined. Often a volume of acid, usually hydrochloric, is selected based on the design engineers' past experience. Because there is little published research in this area and no standard practices, we have chosen not to attempt a more detailed discussion of these uses for acid.

2.3 Theoretical Productivity Improvement From Acidization

Matrix Acidizing

A matrix acid treatment will be effective primarily in a well with a near-wellbore flow restriction, often called a damaged well. To illustrate the productivity improvement possible by removal of near-wellbore damage, consider the simplified radial system illustrated in Fig. 2.1. In this system, a zone of reduced permeability, k_s, extends from the wellbore radius, r_w, to a radius, r_s, beyond which the formation has a constant permeability, k, to the drainage radius,

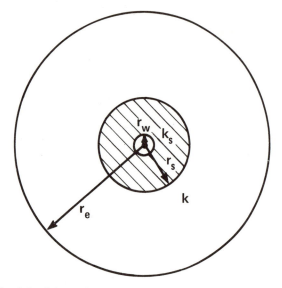

Fig. 2.1—Schematic of a damaged well in a bounded reservoir.

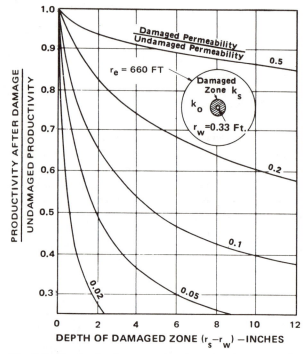

Fig. 2.2—Production loss caused by formation damage (radial flow).

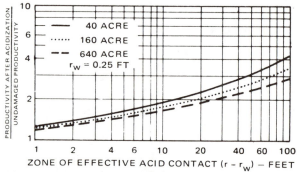

Fig. 2.3—Maximum stimulation ratio attainable in an undamaged well with a matrix acid treatment.

r_e.* Muskat[1] has shown that the fluid production for this system compared with that of a similar system of uniform permeability k is given by

$$\frac{J_s}{J_o} = \frac{F_k \log r_e/r_w}{\log r_s/r_w + F_k \log r_e/r_s} , \ldots\ldots\ldots\ldots (2.1)$$

where $F_k = k_s/k_o$, the permeability ratio, J_o is the undamaged formation productivity, and J_s is the productivity of the damaged well. This expression is plotted in Fig. 2.2 for values of $r_s - r_w$ of 0 to 12 in. for a well with a 660-ft drainage radius.

The production increase that can be obtained from damage removal can be estimated from Fig. 2.2, provided both the radius of the damaged zone and its permeability relative to the formation permeability are known. For example, if the damage zone extends 6 in. into the formation and the permeability ratio is 0.05, then the productivity will only be 0.3 of the productivity of an undamaged well. A stimulation treatment that removes this damage will give a 3.3-fold increase in production rate.

A matrix treatment provides very little stimulation in an undamaged well. For illustration, Fig. 2.3 shows the maximum stimulation attainable in an undamaged well with a matrix treatment. This plot indicates that to achieve a twofold increase in productivity index in a well completed on 40-acre spacing, it is necessary to eliminate essentially all the flow resistance within a radius of 13 ft from the wellbore.

Sandstone formations are often treated with a mixture of hydrofluoric and hydrochloric acids at low injection rates to prevent fracturing. This mixture, chosen because of its ability to dissolve the clays found in drilling mud, also will react with most constituents of naturally occurring sandstones, including silica, feldspar, and calcareous material. Acid penetration of the formation is thought to be, more-or-less, radial and, as a result of its high reaction rate with clays, unspent acid reaches only the first few inches around the wellbore. Even if the reaction rate could be reduced substantially, a large acid volume would be required just to fill formation pore space to the radius required to generate appreciable stimulation. Required volumes are illustrated in Table 2.1 for a formation with 20-percent porosity and a 6-in. wellbore diameter. The acid volumes required just to fill the pore volume to any significant radius are considerably in excess of those commonly employed.

In carbonates, matrix treatments normally employ hydro-

*In this monograph, the drainage radius is defined as one-half the well spacing. An alternative definition, which will give slightly different results, specifies the drainage radius as the radius of the circle with an area equal to one-half the well spacing squared (for uniformly spaced wells).

TABLE 2.1—VOLUME REQUIRED TO FILL FORMATION POROSITY

Radius (ft)	Pore Volume* (gal/ft)
1	4.4
5	117
10	471
25	2,840
50	11,800

*Assumes 20-percent porosity and a 6-in.-diameter wellbore.

chloric acid. The high reaction rate of this acid with limestone or dolomite results in the formation of large flow channels, often called "wormholes," shown schematically in Fig. 2.4. Wormholes originate at a perforation and penetrate into the formation. Data presented in Chapter 10 indicate that wormhole length normally will range from a few inches to a few feet. Because of this limited penetration, matrix treatments in carbonates normally can only bypass near-wellbore flow restrictions and do not create significant stimulation above that achieved from damage removal.

In exceptional cases, matrix treatments can give significant stimulation in undamaged wells. This can occur, for example, in naturally fractured formations where acid can flow along existing fractures. In some instances in fractured carbonates, oil-external emulsified acids have been reported to give two- to threefold stimulation ratios above that expected for damage removal alone.

Acid Fracturing

Acid fracturing is the most widely used acidizing technique for stimulating limestone or dolomite formations. In an acid fracturing treatment, a pad fluid is injected into the formation at a rate higher than the reservoir matrix will accept. This rapid injection produces a buildup in wellbore pressure until it is large enough to overcome compressive earth stresses and the rock's tensile strength. At this pressure the rock fails, allowing a crack (fracture) to be formed. Continued fluid injection increases the fracture's length and width. Acid is then injected into the fracture to react with the formation and create a flow channel that extends deep into the formation and remains open when the well is placed back on production.

Stimulation is achieved either by creating a flow path through a damaged zone around the wellbore or by altering the flow pattern in the reservoir. Small-volume acid treatments can overcome wellbore damage and restore native productivity to a well by removing flow restrictions caused by a zone of low permeability near the wellbore. The steady-state production increase for this type treatment can be estimated from Fig. 2.2 as previously discussed.

The productivity of a well following an acid fracturing treatment is often difficult to predict — particularly in gas reservoirs. Problems occur because flow does not stabilize for a significant period of time in low-permeability reservoirs; fluids and additives used in the fracturing treatment can restrict productivity; and, in some instances, flow of the reservoir fluids along the fracture can be turbulent, thereby reducing fracture flow capacity. Since these effects are more pronounced in gas reservoirs, productivity after acid fracturing in a gas reservoir is emphasized in the remainder of this section.

In general, industry has accepted the McGuire and Sikora[2] correlation shown in Fig. 2.5 as a basis for predicting the stimulation ratio for a fractured well, although similar studies have been reported by Prats,[3] Raymond and Binder,[4] and Tinsley et al.[5] Fig. 2.5 can be used only when the pressure transient introduced with initiation of production has reached the well drainage radius.

Fig. 2.5 shows that after stabilization the important variables are the ratio of fracture length to drainage radius, L/r_e, and the ratio of fracture conductivity to formation permeability, $wk_f h_g/kh_n \sqrt{40/A}$, expressed in inches. If, for example, a treatment generated a highly conductive fracture ($wk_f h_g/kh_n \sqrt{40/A} = 10^4$ in.) 130 ft long in a well with a 660-ft drainage radius ($L/r_e = 130/660 = 0.2$), a stimulation ratio of 3.7 would be expected in an undamaged well once production has stabilized.

The production rate of a well following stimulation is normally not constant but will decrease rapidly until the pressure transient introduced by flow reaches the well drainage radius. Before flow stabilization, the observed stimulation ratio will exceed the stabilized ratio predicted by Fig. 2.5.[6-9] For example, results of calculations made with a reservoir simulator (plotted in Fig. 2.6) show that for a typical 0.1-md formation, the stabilization time is about 1,300 days. In a 10-md formation, flow stabilizes in about 18 days. After stabilization, the stimulation ratio agrees with predictions made with the McGuire and Sikora[2] curves.

The stimulation ratio achieved by the fracturing treatment in a gas well is often smaller than expected in a comparable

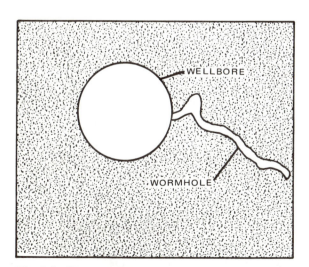

Fig. 2.4—Wormhole pattern for matrix acid treatment in a carbonate formation.

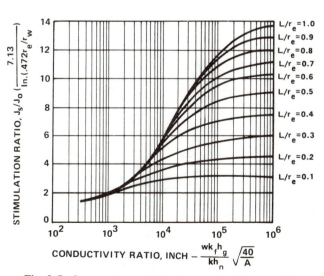

Fig. 2.5—Increase in production from vertical fractures.[2]

oil reservoir because turbulence in the fracture reduces the fracture flow capacity. This effect was discussed by Tannich[10] and Cooke[11] for hydraulic fracturing treatments. A comparable study of turbulence effects in a fracture created by acid reaction has not yet been published. Once data are available, the effect of turbulence in the fracture created by acid reaction should be considered in acid fracturing stimulation-ratio calculations.

If the pad fluid, acid, or additives permanently reduce the formation permeability adjacent to the fracture by a factor of 1,000 or more, the effectiveness of the treatment can be reduced.[3] The effect of a 6-in.-wide damage zone adjacent to a fracture is shown for a typical 0.1-md formation in Fig. 2.7. This figure shows that if the ratio of the damage-zone permeability to the undamaged-formation permeability is 0.01, the production rate will be only slightly reduced. A 10,000-fold reduction in permeability, however, will reduce the well productivity drastically. The damage created adjacent to the fracture is, therefore, seldom great enough to restrict productivity appreciably.

Fluids used in the fracturing treatment can be detrimental to well performance. These fluids can be harmful by making it difficult to get the well on production at a stimulated rate (cleanup problems) or by reducing formation permeability adjacent to the fracture, thereby reducing the effectiveness of the fracture. Tannich[10] has shown that following stimulation, a gas well normally will not be capable of the high rate calculated assuming the fracturing fluid does not impair production. Rather, as shown in Fig. 2.8, the rate will increase with time as the fracturing liquids are removed from the formation and will approach the ideal rate after liquid removal is complete. As fluid viscosity at reservoir conditions increases, this cleanup time also will increase. If the viscosity is extreme, cleanup may in effect never be complete.

The time required to achieve cleanup in an undamaged formation is a function of the formation and fracture permeability, reservoir pressure, viscosity of the fracturing fluid present in the formation, and wellbore configuration (tubing size, well depth, etc.). Theoretical calculations for a typical low-permeability reservoir, given in Fig. 2.9, show that the time required to achieve cleanup will increase as fluid viscosity is increased. In this example, a fluid (either oil or water) that has a viscosity of 0.25 cp at reservoir temperature is removed easily from the formation and the maximum rate is attained after about 3 days of production. The 250-cp fluid is difficult to remove from the formation and only 24 percent of the fracture fluid will have been produced after 400 days of production. Similar increases in cleanup time are seen as fracture length increases.[10]

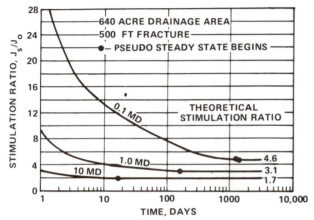

Fig. 2.6—Apparent stimulation ratio obtained by fracturing a homogeneous gas reservoir.

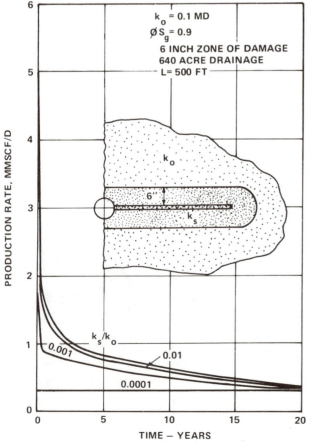

Fig. 2.7—Effect on well productivity of a damage zone adjacent to a fracture.

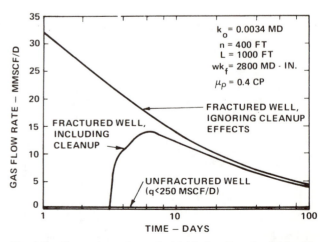

Fig. 2.8—Comparison of actual with ideal well productivities.[10]

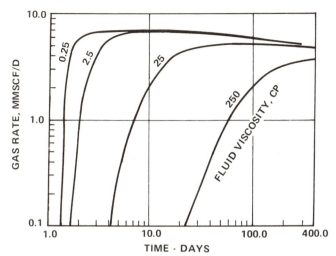

Fig. 2.9—Effect of fracturing fluid viscosity on production rate following stimulation.[10]

A generalized discussion of an acid fracturing treatment is presented in Chapter 5 and a treatment design procedure is given in Chapter 7. We suggest that the engineer interested primarily in acid fracturing treatment design read these two chapters. If more of the theoretical details are desired, Chapter 6 should be consulted.

References

1. Muskat, M.: *Physical Principles of Oil Production,* McGraw-Hill Book Co., Inc., New York (1949) 242.

2. McGuire, W. J. and Sikora, V. J.: "The Effect of Vertical Fractures on Well Productivity," *Trans.,* AIME (1960) **219,** 401-403.

3. Prats, M.: "Effect of Vertical Fractures on Reservoir Behavior — Incompressible Fluid Case," *Soc. Pet. Eng. J.* (June 1961) 105-118; *Trans.,* AIME, **222.**

4. Raymond, L. R. and Binder, G. G.: "Productivity of Wells in Vertically Fractured, Damaged Formations," *J. Pet. Tech.* (Jan. 1967) 120-130; *Trans.,* AIME, **240.**

5. Tinsley, J. M., Williams, J. R., Tiner, L. R., and Malone, W. T.: "Vertical Fracture Height — Its Effect on Steady-State Production Increase," *J. Pet. Tech.* (May 1969) 633-638; *Trans.,* AIME, **246.**

6. Prats, M., Hazebrock, P., and Strickler, W. R.: "Effect of Vertical Fractures on Reservoir Behavior — Compressible Fluid Case," *Soc. Pet. Eng. J.* (June 1962) 87-94; *Trans.,* AIME, **225.**

7. Russell, D. G. and Truitt, N. E.: "Transient Pressure Behavior in Vertically Fractures Reservoirs," *J. Pet. Tech.* (Oct. 1964) 1159-1170; *Trans.,* AIME, **231.**

8. Morse, R. A. and Von Gonten, W. D.: "Productivity of Vertically Fractured Wells Prior to Stabilized Flow," *J. Pet. Tech.* (July 1973) 807-811.

9. Joyner, H. D. and Lovingfoss, W. J.: "Use of a Computer Model in Matching History and Predicting Performance of Low Permeability Gas Wells," *J. Pet. Tech.* (Dec. 1971) 1415-1420.

10. Tannich, J. D.: "Liquid Removal From Hydraulically Fractured Gas Wells," *J. Pet. Tech.* (Nov. 1975) 1309-1317.

11. Cooke, C. E.: "Conductivity of Fracture Proppants in Multiple Layers," *J. Pet. Tech.* (Sept. 1973) 1101-1107; *Trans.,* AIME, **255.**

Chapter 3

Acid Types and the Chemistry of Their Reactions

3.1 Introduction

Practical and economic formation stimulation requires the proper selection of acid type as well as acidizing technique. Essential to this selection is a knowledge of the capabilities and limitations of the acids available. This chapter describes the characteristics of commonly available acids and important chemical factors that should assist in acid selection.

Stoichiometry, equilibrium, and *reaction rate* are the three interrelated but distinct chemical factors that must be considered in selecting the appropriate acid for a particular stimulation treatment.

The *stoichiometry* of an acid reaction with reservoir materials relates the molecular ratio between reactants and reaction products. Once the stoichiometry is known, the quantity of formation material dissolved by a volume of acid can be calculated easily. A useful parameter that relates acid stoichiometry is the *dissolving power.*

A thermodynamic *equilibrium* is established in many acid reactions before the acid has totally reacted. In particular, equilibrium is reached in the reaction of organic acids (such as acetic or formic acid) with limestone and dolomite formations. Equilibrium considerations also control precipitation of reaction products that may negate treatment benefits in either carbonate or sandstone formations.

The *reaction rate* between a particular acid and the formation material fixes the time required for the acid to react. By considering this time along with the geometry within which reaction occurs, the distance acid penetrates away from the well (and thereby the expected stimulation ratio) may be estimated. Because of the importance and complexity of acid reactions, Chapter 4 is devoted to this subject.

3.2 Acid Systems and Considerations in Their Selection

Acid systems in current use can be classified as mineral acids, dilute organic acids, powdered organic acids, hybrid (or mixed) acid systems, or retarded acid systems. The most common members of each category are as follows.

- Mineral acids
    Hydrochloric acid
    Hydrochloric-hydrofluoric acid
- Organic acids
    Formic acid
    Acetic acid
- Powdered acids
    Sulfamic acid
    Chloroacetic acid
- Acid mixtures
    Acetic-hydrochloric acid
    Formic-hydrochloric acid
    Formic-hydrofluoric acid
- Retarded acid systems
    Gelled acids
    Chemically retarded acids
    Emulsified acids

All these, with the exception of the hydrochloric-hydrofluoric and formic-hydrofluoric acid mixtures, are used to stimulate carbonate formations. The characteristics and principal applications of each follow.

Mineral Acids

Hydrochloric Acid — Most acid treatments of carbonaceous formations overwhelmingly employ hydrochloric acid. Usually, it is used as a 15-percent (by weight) solution of hydrogen chloride gas in water. This concentration, often called regular acid, was originally chosen because of inadequacies in early inhibitors and the difficulty of preventing corrosion of well tubulars by more concentrated solutions. Recently, with the development of improved inhibitors, higher concentrations have become practical and, in some cases, they provide increased effectiveness.

In addition to concentrations higher than 15 percent, lower concentrations are commonly available and are used where acid dissolving power is not the sole consideration. An example of such an application is found in sandstone acidizing where 5- to 7.5-percent HCl is often used to displace connate water ahead of hydrochloric-hydrofluoric acid mixtures to prevent the formation of sodium and potassium fluosilicates — materials capable of plugging the formation.

The continued use of hydrochloric acid results from its moderate cost and soluble reaction products (calcium chloride and carbon dioxide). The economy of hydrochloric acid compared with other acidizing materials is demonstrated in Section 3.3 in terms of the dissolving power of the acid.

The principal disadvantage of hydrochloric acid is its high corrosivity on wellbore tubular goods. This high corrosivity is especially significant and expensive to control at temperatures above 250 °F. Also, aluminum- or chromium-plated metals, often found in pumps, are easily damaged. Generally, the application (because of formation temperature or material to be protected) will dictate whether a less corrosive acid than hydrochloric is required.

Hydrochloric-Hydrofluoric Acid — This acid mixture is used almost exclusively for sandstone stimulation. Within the chemical industry, hydrofluoric acid (HF) is available commercially as a relatively pure material in anhydrous form or as a concentrated (40 to 70 percent) aqueous solution. As it is used in the petroleum industry for well stimulation, HF is most often a dilute solution in hydrochloric acid (HCl). It may be formed from dilution of concentrated solutions of hydrogen fluoride or, more frequently, from the reaction of ammonium bifluoride with hydrochloric acid. Often, 15-percent HCl is used, and enough ammonium bifluoride is added to create a solution containing 3-percent HF. Consumption of hydrogen chloride by this reaction leaves 12-percent HCl remaining in solution. Similarly, 6-percent HF is often generated from 15-percent HCl solutions and the final hydrochloric acid concentration is approximately 9 percent. More concentrated solutions ranging from 12-percent HCl and 10-percent HF to 25-percent HCl and 20-percent HF are prepared by dilution of high-strength HF with an aqueous HCl solution. The corrosion characteristics of the HF-HCl mixture are comparable with those of HCl alone, and similar corrosion inhibitors are required.

Organic Acids

The principal virtues of the organic acids are their lower corrosivity and easier inhibition at high temperatures. They have been used primarily in operations requiring a long acid-pipe contact time, such as a perforating fluid, or where aluminum or chrome-plated parts unavoidably will be contacted. Although many organic acids are readily available, only two, acetic and formic, are used to any great extent in well stimulation.

Acetic Acid — Acetic acid was the first of the organic acids to be used in appreciable volumes in well stimulation. It is commonly available as a 10-weight-percent solution of acetic acid in water. At this concentration, the products of reaction (calcium and magnesium acetates) are generally soluble in spent acid. In addition to being used as a perforating fluid or as a fluid of low corrosivity in the presence of metals that corrode easily, acetic acid is often used in mixture with hydrochloric acid in the hybrid acids.

On the basis of cost per unit of dissolving power, acetic acid is more expensive than either hydrochloric or formic acids. This greater expense generally dictates its use in small quantities only for the special applications described.

Acetic acid reacts incompletely in the presence of its reaction products. The equilibrium set up between the products and reactants of the system often has been misinterpreted as evidence of a retarded reaction rate. Because of

this, organic acids sometimes have been sold as retarded acids for regular acidizing operations. A discussion of acid equilibrium reactions is given later in this chapter.

Formic Acid — Of the organic acids used in acidization, formic acid has the lowest molecular weight and, correspondingly, the lowest cost per volume of rock dissolved. It is substantially stronger than acetic acid, though appreciably weaker than hydrochloric acid. Like acetic acid, it reacts to an equilibrium concentration in the presence of its reaction products.

The principal advantage of formic over acetic acid is cost, although this is partially offset by the greater difficulty of inhibiting corrosion with this acid. Although more corrosive than acetic acid, formic acid corrodes uniformly and with less pitting than hydrochloric acid, and effective inhibitors are available for its use at temperatures as high as 400 °F. In high-temperature applications, the cost discrepancy with hydrochloric acid narrows because of the high inhibitor concentrations required for HCl.

Powdered Acids

Sulfamic and Chloroacetic Acids — These two acids have only limited use in well stimulation, most of which is associated with their portability to remote locations in powdered form. They are white crystalline powders that are readily soluble in water. Generally, they are mixed with water at or near the wellsite. Sometimes these acids are cast in useful shapes such as ''acid sticks,'' a form convenient for introduction into the wellbore.

Both sulfamic and chloroacetic acids are substantially more expensive than hydrochloric acid on an equivalent-dissolving-power basis. Significant savings accompany their use when transportation and pumping charges can be eliminated.

Chloroacetic acid is stronger and more stable than sulfamic acid and is generally preferred when a powdered acid is appropriate. Sulfamic acid decomposes at about 180 °F and is not recommended for applications where formation temperatures are above 160 °F.

Acid Mixtures

Acetic-Hydrochloric and Formic-Hydrochloric Acids — These acid mixtures, useful on carbonates, generally have been designed to exploit the dissolving-power economies of hydrochloric acid while attaining the lower corrosivity (especially at high temperatures) of the organic acids. Therefore, their application is almost exclusively in high-temperature formations where corrosion inhibition costs greatly affect the over-all treatment cost. The mixed acids sometimes have been sold as retarded acids because of the presence of the organic acid. It is important to recognize that, under formation conditions, CO_2 evolved by the HCl reaction greatly reduces the extent of reaction of organic acids and, in extreme cases, may prevent the organic acid from reacting.

Formic-Hydrofluoric Acid — This acid mixture, useful on sandstones, is sometimes employed in high-temperature applications because it is less corrosive than the comparable inorganic acid mixture, HF-HCl.

Retarded Acid Systems

The acid reaction rate theoretically can be retarded by gelling the acid, oil-wetting the formation solids, or emulsifying the acid with an oil.

Gelled Acids — Gelled acids are used to retard acid reaction rate in fracturing treatments. Retardation results because the increased fluid viscosity reduces the rate of acid transfer to the fracture wall. Use of the gelling agents (normally water-soluble polymers) is limited to low-temperature formations because most of the available agents degrade rapidly in acid solution at temperatures exceeding about 130 °F. When more stable polymers are developed, they should find application in acid fracturing. Gelling agents are seldom used in matrix acidization because the increased acid viscosity reduces injectivity of the acid and often prolongs the treatment needlessly.

Chemically Retarded Acids — These acids are often prepared by adding an oil-wetting surfactant to acid in an effort to create a physical barrier to acid transfer to the rock surface. To function, the additive must adsorb on the rock surface and form a coherent film. Use of these acids often requires continuous injection of oil during the treatment. At high flow rates and high formation temperatures, adsorption is diminished and most of these materials become ineffective.

Emulsified Acids — Emulsified acids may contain the acid as either the internal or the external phase. The former, which is more common, normally contains 10 to 30 percent hydrocarbon as the external phase and regular hydrochloric acid as the internal phase. When acid is the external phase, the ratio of oil to acid is often about 2:1. Both the higher viscosity created by emulsification and the presence of the oil can retard the rate of acid transfer to the rock. This reduction in transfer rate, and its corresponding reduction in acid reaction rate, often can increase the depth of acid penetration. Use of oil-external emulsified acids is occasionally limited by the increased frictional resistance to flow of these fluids down well tubulars.

3.3 Stoichiometry of Acid Carbonate Reactions

Stoichiometry refers to the proportions of the various reactants that participate in the reaction. While these proportions are easy to identify in the reaction of calcium carbonate or dolomite with hydrochloric acid, reactions with naturally occurring carbonates are often complicated by the presence of other minerals reactive with HCl.

Chemical Composition of Carbonate Formations

Carbonate rocks have been created either by chemical and biochemical precipitation in a water environment or by transportation of clastic grains. The active process is often difficult to determine because a clastic limestone may be so thoroughly recemented and recrystallized as to be readily mistaken for a precipitate. The chemically precipitated carbonate reservoir rocks are usually crystalline limestones and dolomites, but marl and chalk are also in this category. Some carbonate rocks are relatively pure while others contain siliceous materials. The siliceous components may be

precipitated chert, siliceous fossils, clastic grains of quartz or chert, or shaly material. As the concentration of siliceous components increases, rocks commonly classed as sandy, cherty, or shaly limestones or dolomites are formed.

The carbonates originally precipitated in a rock may be almost pure calcite ($CaCO_3$) or dolomite [$CaMg(CO_3)_2$]. These two carbonates, however, are often intermingled. Dolomitic and calcitic rocks may be interbedded, or the calcium in a calcite may have been partially replaced by magnesium, and the rock thereby formed is classed as a magnesian or dolomitic limestone. Similarly, it may be regarded as a calcareous dolomite if it contains more than 50 percent dolomite. The relationship between rock composition and geologic classification is given in Fig. 3.1. (A more detailed discussion of the classification of carbonate rocks is given in Refs. 1 through 3.)

The remainder of this chapter is directed toward pure limestones or dolomites. Conclusions may be applied to mixed carbonates once the composition of the reservoir rock is known.

Acid Reactions With Carbonates

The acids commonly used to treat carbonate formations are listed in Table 3.1. These acids all react with carbonates to form carbon dioxide (CO_2), water, and a calcium or magnesium salt. Typical reactions are

$$2HCl + CaCO_3 \underset{\leftarrow}{\rightarrow} CaCl_2 + H_2O + CO_2, \quad \ldots\ldots\ldots \text{ (3.1)}$$

and

$$4HCl + CaMg(CO_3)_2 \underset{\leftarrow}{\rightarrow} CaCl_2 + MgCl_2$$
$$+ 2H_2O + 2CO_2. \quad \ldots\ldots\ldots\ldots\ldots\ldots \text{ (3.2)}$$

These equations indicate the stoichiometry of the reaction. For example, Eq. 3.1 indicates that 2 moles of hydrochloric acid (HCl) react with 1 mole of limestone (calcium carbonate, $CaCO_3$) to create 1 mole of calcium chloride ($CaCl_2$), 1 mole of water (H_2O), and 1 mole of carbon dioxide (CO_2).

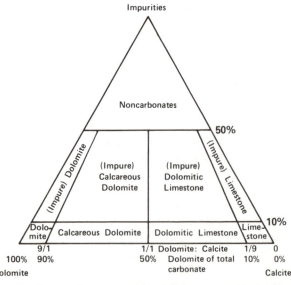

Fig. 3.1—Compositional graph showing the relationship between the carbonate rocks.[1]

The numbers multiplying the moles of the component required in the reaction (for example, "2"HCl) are known as *stoichiometric coefficients*. Combining Eq. 3.1 with molecular weight data for each component (given in Table 3.2) allows calculation of the amount of acid required to dissolve a given quantity of a carbonate, the quantity of reaction products produced by the reaction, or other stoichiometric data.

The concept of acid dissolving power (expressed as volume of rock dissolved per unit volume of acid reacted) is a useful quantity because it allows a direct comparison of acid costs. Dissolving power is easily calculated for any reaction of interest:

First, define β to be the mass of rock dissolved per unit mass of acid reacted. Therefore,

$$\beta = \frac{\text{molecular weight of mineral (rock)} \times \text{its stoichiometric coefficient}}{\text{molecular weight of acid} \times \text{its stoichiometric coefficient}} \quad \ldots \ldots (3.3)$$

For the reaction of 100-percent hydrochloric acid with pure limestone, defined by Eq. 3.1, the dissolving power is

$$\beta_{100} = \frac{100.09 \times 1}{36.47 \times 2} = 1.372 \ \frac{\text{gm limestone dissolved}}{\text{gm 100-percent HCl reacted}}.$$

$$\ldots \ldots \ldots \ldots \ldots \ldots \ldots \ldots \ldots \ldots (3.4)$$

If the acid concentration is 15 percent by weight rather than 100 percent, then

$$\beta_{15} = \beta_{100} \times 0.15 = 0.206 \ \frac{\text{gm limestone dissolved}}{\text{gm 15-percent HCl reacted}}.$$

$$\ldots \ldots \ldots \ldots \ldots \ldots \ldots \ldots \ldots \ldots (3.5)$$

The dissolving power, which is the volume of rock dissolved per volume of acid reacted (defined as X) can be obtained from Eq. 3.5 by multiplying the mass ratio by the appropriate density ratio. Note that the porosity of the rock is not included in this calculation. For 15-weight-percent HCl, this calculation gives

$$X_{15} = \frac{\rho_{\text{15-percent HCl}} \, \beta_{\text{15-percent HCl}}}{\rho_{\text{CaCO}_3}}, \quad \ldots \ldots \ldots \ldots (3.6)$$

where $\rho_{\text{15-percent HCl}}$ is the density of a 15-percent-HCl solu-

tion (1.07 gm/cc) and ρ_{CaCO_3} is the density of calcium carbonate (2.71 gm/cc). The specific gravity data for HCl solutions are given in Table 3.3. Substituting into Eq. 3.6 gives

$$X_{15} = \frac{1.07(0.206)}{2.71} = 0.082 \ \frac{\text{cc limestone dissolved}}{\text{cc 15-percent HCl reacted}}.$$

$$\ldots \ldots \ldots \ldots \ldots \ldots \ldots \ldots \ldots \ldots \ldots (3.7)$$

Although the volume units shown in Eq. 3.7 are cubic centimeters, this volumetric ratio is independent of units, and any consistent set of volumetric units may be used. Values for the dissolving power are given in Table 3.4 for hydrochloric acid and the commonly used organic acids. Data are included for several acid concentrations and for both limestone and dolomite formations.

Table 3.4 is useful when comparing one acid with another. In general, HCl has the largest dissolving power, followed by formic acid and then acetic acid. The numbers presented in this table do not take into account limitations that may be imposed by chemical equilibrium. Typically, in field treatments, organic acids do not react completely, so a given volume of the acid will dissolve less rock than is indicated in Table 3.4. To correct the dissolving power, it must be multiplied by the fraction of acid that reacts before

TABLE 3.1—ACIDS USED IN CARBONATE ACIDIZATION

Mineral Acids	Molecular Weight
Hydrochloric (HCl)	36.47
Organic Acids	
Formic (HCOOH)	46.03
Acetic (CH₃COOH)	60.05

TABLE 3.2—MOLECULAR WEIGHT OF COMPONENTS IN HCl REACTION WITH CARBONATES

Compound	Chemical Formula	Molecular Weight
Hydrochloric acid	HCl	36.47
Calcium carbonate (limestone)	CaCO₃	100.09
Calcium magnesium carbonate (dolomite)	CaMg(CO₃)₂	184.3
Calcium chloride	CaCl₂	110.99
Magnesium chloride	MgCl₂	95.3
Carbon dioxide	CO₂	44.01
Water	H₂O	18.02

TABLE 3.3—SPECIFIC GRAVITY OF AQUEOUS HYDROCHLORIC ACID SOLUTIONS[10] (at 20 °C)

Percent HCl	Specific Gravity
1	1.0032
2	1.0082
4	1.0181
6	1.0279
8	1.0376
10	1.0474
12	1.0574
14	1.0675
16	1.0776
18	1.0878
20	1.0980
22	1.1083
24	1.1187
26	1.1290
28	1.1392
30	1.1493
32	1.1593
34	1.1691
36	1.1789
38	1.1885
40	1.1980

TABLE 3.4—DISSOLVING POWER OF VARIOUS ACIDS*

	β_{100}**	X† 5 Percent	X† 10 Percent	X† 15 Percent	X† 30 Percent
Limestone (CaCO₃, calcite: ρ_{CaCO_3} = 2.71 gm/cc)					
Hydrochloric (HCl)	1.37	0.026	0.053	0.082	0.175
Formic (HCOOH)	1.09	0.020	0.041	0.062	0.129
Acetic (CH₃COOH)	0.83	0.016	0.031	0.047	0.096
Dolomite [CaMg(CO₃)₂: $\rho_{\text{CaMg(CO}_3)_2}$ = 2.87 gm/cc]					
Hydrochloric	1.27	0.023	0.046	0.071	0.152
Formic	1.00	0.018	0.036	0.054	0.112
Acetic	0.77	0.014	0.027	0.041	0.083

*Data for organic acids have not been corrected for equilibrium.

$$**\beta_{100} = \frac{\text{mass rock dissolved}}{\text{mass pure acid reacted}}.$$

$$\dagger X = \frac{\text{volume rock dissolved}}{\text{volume acid solution reacted}}.$$

equilibrium at the reaction conditions (formation temperature, pressure, and concentration of products present). Procedures for estimating the extent of acid reaction are presented in the following section.

3.4 Equilibrium In Acid-Carbonate Reactions

Along with the concepts of reaction stoichiometry and rate, it is important to understand *equilibrium* as it relates to acid reactions. When an acid reaction reaches equilibrium, the dissolution of formation material by the acid stops, even though acid molecules still may be present. Equilibrium is attained when the chemical activity (consider this to be a driving force for change) of the reaction products balances the activity of the reactants.

A general definition of equilibrium can be obtained from thermodynamic arguments.[4] The final result is that at equilibrium, the ratio of the chemical activity of products to reactants, with each activity raised to a power equal to its stoichiometric coefficient, is equal to a constant, called an *equilibrium constant*. For example, consider the generalized reaction

$$A + B \underset{\leftarrow}{\rightarrow} C + D.$$

The equilibrium constant for this reaction can be defined as

$$K = \frac{a_C a_D}{a_A a_B} . \quad\dots\dots\dots\dots\dots\dots\dots\dots (3.8)$$

The quantity a_i is the activity of the component i. These activities are thermodynamic potentials and are not easy to predict. Therefore, experimental data are generally required to obtain accurate values. The activity of a substance increases with its concentration in solution, but the relationship between activity and concentration is generally not linear. This relationship is often expressed by defining a proportionality constant between the chemical activity and the concentration, called the activity coefficient ($a_i = \gamma_i c_i$). Example activity coefficient data for HCl are given in Table 3.5. More detailed discussions of activities are given in texts on chemical thermodynamics.[4, 5]

Dissociation Equilibrium

An important property of acids is that, in aqueous solution, they dissociate (ionize) by the reaction

$$HA \rightleftharpoons H^+ + A^-. \quad\dots\dots\dots\dots\dots\dots\dots (3.9)$$

In this equation, the generalized acid is denoted as HA with the ionized species being H^+ and A^- ions. For example, hydrogen chloride ionizes to produce hydrogen ions (H^+) and chloride ions (Cl^-). Equilibrium for hydrochloric acid dissociation is described by

$$K_D = \frac{a_{H^+} a_{Cl^-}}{a_{HCl}} . \quad\dots\dots\dots\dots\dots\dots\dots (3.10)$$

The equilibrium constant, K_D, in this case is called the *dissociation constant*. If at equilibrium an acid is highly dissociated, K_D will be a large number; if the acid is only slightly dissociated, K_D will be small.

The dissociation constants depend on temperature and can be calculated using Eq. 3.11.

$$-\log_{10} K_D = \frac{A_1}{T} - A_2 + A_3 T. \quad\dots\dots\dots\dots (3.11)$$

In this equation, T is the temperature in degrees Kelvin and the constants A_1, A_2, and A_3 can be obtained from Table 3.6. Values of K_D for acids used in carbonate formations are summarized in Table 3.7.

As seen from data in Table 3.7, acetic and formic acids have small dissociation constants compared with hydrochloric acid. This means that under comparable conditions a much smaller quantity of the acid will be dissociated into the reactive ionized state. Therefore, they are often called *weak acids*.

Reaction Equilibrium

Under reservoir conditions, organic acids do not react to completion with either limestone or dolomite formations because of the limitations imposed by chemical equilibrium. Equilibrium occurs in the reservoir because CO_2 (one reaction product) is held in solution by reservoir pressure and not allowed to escape from the solution.[8] At low pressures, where the CO_2 can escape, the acid will, however, react to completion.

Results of tests relating the fraction of acid reacted, temperature, and acid composition at 1,000 psi are given in Figs. 3.2 and 3.3.[8] These data are consistent with those presented by Harris[9] and, since all CO_2 is in solution at 1,000 psi, should be applicable at all higher pressures. These factors can be used to correct the dissolving power of organic acids. For example, Fig. 3.2 shows that at 150 °F and 1,000 psi, only about 50 percent of a 10-weight-percent acetic acid solution will react. The dissolving power of acetic acid under these conditions therefore should be reduced by 50 percent from the value given in Table 3.4.

For conditions not considered in Figs. 3.2 and 3.3, we have found that the equilibrium state can be predicted *approximately* by the empirical equation

$$1.6 \times 10^4 K_D = \frac{c_{CaA_2} c_{CO_2}}{c_{HA}} , \quad\dots\dots\dots\dots (3.12)$$

where c_i is the concentration of component i in gram-moles per 1,000 gm of water. Experimental evidence indicates that equilibrium conditions with dolomite are very nearly the same as limestone; therefore, Eq. 3.12 also should apply to

TABLE 3.5—ACTIVITY COEFFICIENT OF HYDROCHLORIC ACID[6]

Concentration (mol/liter)	Activity Coefficient
0.1	0.80
0.5	0.76
1.0	0.81
2.0	1.04
4.0	1.96
6.0	4.19
8.0	9.60
12.0	32.16

TABLE 3.6—CONSTANTS FOR DETERMINING ACID DISSOCIATION CONSTANTS[7]

Acid	A_1	A_2	A_3
Acetic	1,170.48	3.1649	0.013399
Formic	1,342.85	5.2743	0.015168
Propionic	1,213.26	3.3860	0.014055
Chloroacetic	1,229.13	6.1714	0.016486

TABLE 3.7—TYPICAL VALUES FOR ACID DISSOCIATION CONSTANTS[7]

	Dissociation Constant, K_D			
	77 °F	100 °F	150 °F	250 °F
K_D (acetic)	1.754×10^{-5}	1.716×10^{-5}	1.4822×10^{-5}	8.194×10^{-6}
K_D (formic)	1.772×10^{-4}	1.735×10^{-4}	1.486×10^{-4}	7.732×10^{-5}
K_D (HCl)	10			

dolomites.[8] An application of this calculation for a limestone follows.

Example — Calculation of Equilibrium

Consider the reaction of 10-weight-percent acetic acid with limestone at 150 °F and 1,000 psi.

1. Determine the concentration of acetic acid in gram-moles per 1,000 gm of water, c_{HAc}. In 1,000 gm of solution, there are 100 gm of acid and 900 gm of water. Therefore,

$$\frac{\text{mol HAc}}{1,000 \text{ gm solution}} = \frac{100 \text{ gm HAc}}{60 \left(\dfrac{\text{gm HAc}}{\text{gm-mol HAc}} \right)} = 1.67,$$

$$\dots\dots\dots\dots\dots\dots\dots\dots\dots\dots (3.13)$$

$$c_{HAc} = \frac{1.67 \text{ mol HAc}}{1,000 \text{ gm solution}} \cdot \frac{1,000 \text{ gm solution}}{900 \text{ gm water}}$$

$$= \frac{1.85 \text{ mol}}{1,000 \text{ gm H}_2\text{O}}, \dots\dots\dots\dots\dots (3.14)$$

where the molecular weight of HAc was obtained from Table 3.1.

2. Calculate the equilibrium composition from Eq. 3.12.

If ψ is the number of moles of acetic acid per 1,000 gm of water that react, then at equilibrium*

$$CaCO_3 + 2H(CH_3COO) \rightleftarrows Ca(CH_3COO)_2 + H_2O + CO_2,$$

$$c_{CO_2} = \psi/2,$$

$$c_{Ca(CH_3COO)_2} = \psi/2, \dots\dots\dots\dots (3.15)$$

$$c_{H(CH_3COO)} = 1.85 - \psi \text{ (taking the initial value for}$$
$$c_{H(CH_3COO)} = 1.85 \text{ from}$$
$$\text{Eq. 3.14).}$$

The value of K_D is 1.48×10^{-5} (found using either Eq. 3.11 with Table 3.6 or Table 3.7). Substituting directly into Eq. 3.12 gives

$$(1.6 \times 10^4) (1.48 \times 10^{-5}) = \frac{(\psi/2)(\psi/2)}{1.85 - \psi}.$$

Solving the quadratic for ψ yields an amount reacted of 0.93 mol per 1,000 gm of water.

The fraction of acid converted is then $0.93/1.85 = 0.50$. The experimental value for these conditions from Fig. 3.2 is 0.54. As illustrated, the prediction obtained using Eq. 3.12 normally is adequate for most practical purposes.

This use of Eq. 3.12 assumes that both the CO_2 and the calcium-acid anion salt remain in the aqueous solution. Although the reservoir pressures are generally sufficient to force the CO_2 to remain in solution when modest acid

concentrations are used, a separate CO_2-rich phase may form if the initial concentration of the acid is 15 percent or greater. A second possibility is that CO_2 may escape from the aqueous phase into an intermingled oil phase if one is present. In either case, the concentration of CO_2 to use in Eq. 3.12 is that actually remaining in the aqueous phase. Clearly, reducing the concentration of CO_2 in the aqueous phase shifts the reaction so that more acid reacts.

It is useful to note again that the reaction of an organic acid in a hybrid acid mixture is probably subject to an extreme thermodynamic limitation since the HCl will generally react first, producing a large quantity of CO_2. In accordance with Eq. 3.12, the added CO_2 will reduce the quantity

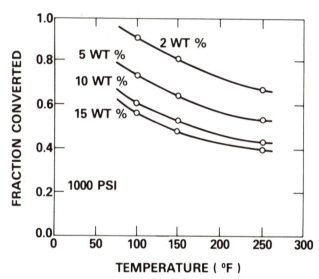

Fig. 3.2—Fraction acid reacted at equilibrium vs temperature for the system acetic-acid/calcium-carbonate.[8]

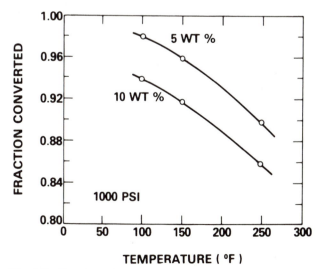

Fig. 3.3—Fraction acid reacted at equilibrium vs temperature for the system formic-acid/calcium-carbonate.[8]

---

*This calculation ignores the dilution effect of the small quantity of water produced by the reaction. This is generally a reasonable assumption for practical application.

of organic acid that can react before Eq. 3.12 is satisfied. Therefore, there is likely to be more unreacted organic acid than is indicated by Figs. 3.2 and 3.3 when hybrid acids are used.

The solubility of CO_2 in a spent acid solution has not been measured, but the solubility in water, given in Table 3.8, can be used to approximate solubility values needed for use in equilibrium calculations.

If the CO_2 concentration is predicted to exceed that given in Table 3.8, the solubility from the table should be used in Eq. 3.12. This consideration is often important at low pressures, for high concentrations of organic acids, for hybrid acids, or when the formation contains appreciable CO_2.

The solubilities of calcium salts in water are given in Table 3.9. If the predicted salt concentration exceeds the value shown, the solubility from the table should be used in Eq. 3.12. Perhaps of even greater practical significance than the equilibrium conversion is the possible plugging of the formation with the precipitated salt from the organic acid if the salt concentration exceeds the solubility limit.

3.5 Stoichiometry of Acid-Sandstone Reactions

Acidizing treatments in sandstone formations normally employ a mixture of hydrochloric and hydrofluoric acids. This acid mixture is used because hydrofluoric acid is reactive with clay minerals that may be restricting near-wellbore permeability. These permeability restrictions can be caused by clays introduced into the formation from the drilling mud or by altering clays present in the formation if mud filtrate changes their chemical environment. Hydrochloric acid alone is sometimes used to treat sandstones, but these treatments are normally successful only when the sandstone has a high calcite (calcium carbonate) content.

Chemical reactions between hydrofluoric acid and silica or calcite in the rock matrix are comparatively simple. The reactions of hydrofluoric acid with silicates such as clays or feldspars are complex, however, because these minerals, which occur as three-dimensional lattices, may not be represented by a single stoichiometric expression. Empirical formulas are often employed to represent the average ratios of their constituent elements. Examples for the more common clay minerals are

kaolinite,

$(Al_{1.8}Fe_{0.1}{}^{+3}Mg_{0.1})Si_2O_5(OH)_4 Ca_{0.05}$,

montmorillonite,

$Al_{1.67}Mg_{0.33}Si_4O_{10}(OH)_2Na_{0.33}$,

and feldspars such as

albite,

$(NaSi_3Al)O_8$.

Equations that describe the stoichiometry of the hydrofluoric acid reaction with silica, the silicate minerals, and calcite follow. In these equations, the reaction of hydrofluoric acid with sodium silicate is used to represent the reaction with silicates found in the sandstone matrix.

Reaction With Silica

$$SiO_2 + 4HF \rightleftharpoons SiF_4 + 2H_2O. \dots\dots\dots\dots (3.16)$$

$$SiF_4 + 2HF \rightleftharpoons H_2SiF_6. \dots\dots\dots\dots (3.17)$$

Reaction With Silicates (Feldspar or Clays)

$$Na_4SiO_4 + 8HF \rightleftharpoons SiF_4 + 4NaF + 4H_2O. \dots (3.18)$$

$$2NaF + SiF_4 \rightleftharpoons Na_2SiF_6. \dots\dots\dots\dots (3.19)$$

$$2HF + SiF_4 \rightleftharpoons H_2SiF_6. \dots\dots\dots\dots (3.20)$$

Reaction With Calcite

$$CaCO_3 + 2HF \rightleftharpoons CaF_2 + H_2O + CO_2. \dots\dots (3.21)$$

Based on these reactions, the dissolving power of hydrofluoric acid can be computed as was done for carbonates. The results are given in Table 3.10. Note that as previously discussed, the HCl is *not appreciably reactive with sand and clay minerals* and therefore is not included in the dissolving-power calculation.

The mineralogical content of oil-bearing formations varies widely, as illustrated in Table 3.11. The volume of formation that will be dissolved by a volume of HF-HCl acid varies correspondingly. While in principal it is possible to compute the acid-soluble portion from its composition,

TABLE 3.8—SOLUBILITY OF CARBON DIOXIDE IN WATER[11]

Pressure (psia)	Moles of CO_2 Per 1,000 gm of Water		
	95 °F	167 °F	212 °F
367.5	0.58	0.31	0.24
735	1.00	0.57	0.46
1,102.5	1.25	0.77	0.64
1,470	1.31	0.92	0.79
1,837.5	1.37	1.10	1.02
2,205	1.43	1.20	1.15
2,940	—	1.32	1.33
4,410	1.60	1.43	1.45
5,880	1.70	—	—
7,350	—	1.69	1.73

TABLE 3.9—SOLUBILITY, IN WATER, OF CALCIUM SALTS OF ORGANIC ACIDS[6]

Temperature (°F)	Grams Per 100 gm of Water	
	Calcium Formate	Calcium Acetate
32	16.15	37.4
50		36.0
68	16.60	34.7
86		33.8
104	17.05	33.2
140	17.50	32.7
176	17.95	33.5
194		31.1
211	18.40	29.7

TABLE 3.10—DISSOLVING POWER FOR HYDROFLUORIC ACID

Acid Concentration (weight-percent)		X, Volume of Rock Dissolved/ Volume of Acid Solution Reacted	
HF	HCl	Clay (Na_4SiO_4)	Sand (SiO_2)
2.1	12.9	0.017	0.007
3.0	12.0	0.024	0.010
4.2	10.8	0.033	0.014
6.0	9.0	0.047	0.020

TABLE 3.11—X-RAY ANALYSES OF TYPICAL GULF COAST SANDSTONE FORMATIONS, DRILLING MUDS, AND COMMERCIAL CLAYS[12]

Field	Sandstone	Depth (ft)	Composition (percent by weight)				
			Quartz	Clay*	Albite	Pyrite	Miscellaneous*
Barbers	Oligocene	—	70	151	15		
Banada	Oligocene	5,100	60 to 70	20 to 40K			
Golden Meadows	Miocene	5,300	60 to 70	20 to 30B	10		
Lake St. John	Wilcox	3,398	90	10M			
		3,438	80	15 M	5		
Loma Novia	Government wells	2,732	45	45 MM	10		
		2,750	50	15MM	30		5C
Manvel	Marginulina	5,355	60 to 70	20 to 30B	5 to 10	1	
		5,375	50 to 60	30 to 40B	5 to 10		
Segno	Wilcox	8,040	60 to 80	10 to 20MM		5 to 10	5 to 10X
		8,030	70 to 80	10 to 20MM			5 to 10X
Sheridan	Wilcox	9,384	50 to 65	15 to 25X	20 to 30		
		9,405	65 to 80	10 to 15H	10 to 20		
Villa Platte	Wilcox	9,844	70 to 90	10 to 20H	5 to 10		
		10,182	65 to 85	10 to 20H	5 to 10	3 to 5	
		10,207	60 to 80	20 to 30H	5 to 10	1 to 3	
	Haas	9,014	70 to 90	10 to 20M	5 to 10		
Wharton	Kountz	5,158	90	10K			
	Lancaster	5,140	50	50K			
Wildcat	Eocene	—	50 to 75	20 to 30MM	10 to 20		

* B—Beidellite   H—Metahalloysite   K—Kaolin   MM—Montmorillonite
C—Calcite   I—Illite   M—Muscovite   X—Unidentified

practically it is often more reliably determined from laboratory tests.

3.6 Equilibrium in Acid-Sandstone Reactions

Hydrofluoric-acid/hydrochloric-acid mixtures are subject to many of the same equilibrium considerations as acids used in carbonate formations. Hydrochloric acid is a strong acid and is essentially totally dissociated into hydrogen and chloride ions when in solution. In the presence of hydrochloric acid, hydrofluoric acid is poorly dissociated and behaves as a weak acid. Also, HF can combine in several chemical states to form complex ion configurations as illustrated by the formation of HF_2^-:

$$HF \underset{\longleftarrow}{\longrightarrow} H^+ + F^-,$$

$$HF + F^- \underset{\longleftarrow}{\longrightarrow} HF_2^-. \quad \dots\dots\dots\dots\dots (3.22)$$

Although the formation of HF_2^- and other equilibrium states are of theoretical interest, in practical application the concentration of these components is very low and they need not be considered. Reaction with sand or clays is thought to be primarily by the undissociated acid molecule, HF.

As shown in Eqs. 3.16 and 3.17, reaction between hydrofluoric acid (HF) and silicon dioxide (SiO_2, sand) produces silicon tetrafluoride (SiF_4). Equilibrium considerations require, however, that very little SiF_4 exist in solution since it is converted very quickly to the fluosilicate ion, as shown below.

$$SiF_4 + 2F^- \underset{\longleftarrow}{\longrightarrow} SiF_6^=. \quad \dots\dots\dots\dots\dots (3.23)$$

In the presence of sodium or potassium, the $SiF_6^=$ can react to form insoluble fluosilicate salts (Na_2SiF_6), or, in their absence, fluosilicic acid (H_2SiF_6):

$$SiF_6^= + 2Na^+ \underset{\longleftarrow}{\longrightarrow} Na_2SiF_6. \quad \dots\dots\dots\dots (3.24)$$

$$SiF_6^= + 2H^+ \underset{\longleftarrow}{\longrightarrow} H_2SiF_6. \quad \dots\dots\dots\dots (3.25)$$

The fluosilicic acid then can be hydrolyzed to orthosilicic

acid $[Si(OH)_4]$ as follows.

$$H_2SiF_6 + 4H_2O \underset{\longleftarrow}{\longrightarrow} Si(OH)_4 + 6H^+ + 6F^-. \quad \dots\dots (3.26)$$

Labrid,[13] in a recent study of the complex equilibrium involved in sandstone acidizing, reached the following conclusion:

"A thermodynamic study of the solubilization process of minerals by hydrofluoric acid reveals that, for silica, the main reaction product of the reaction is fluosilicic acid accompanied by a small amount of colloidal silica. Feldspar and clay solubilization takes place, in the first stage, by a uniform alteration of the crystalline lattice, then by a progressive extraction of aluminum in the form of fluorinated complexes."

At least seven fluorine complexes of aluminum, ranging from $Al_2F_6^=$ to AlF^{++}, are thought to be formed by the contact of hydrofluoric acid with clay minerals. High acid strength originally creates compounds rich in fluorine while, as the reactant depletes, the complexes become lean in fluorine. Fig. 3.4, from Labrid's work, shows the range of fluoride ion concentration in which the different complexes

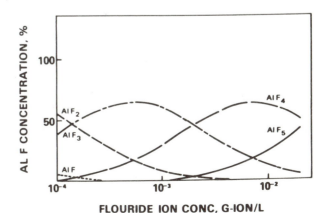

Fig. 3.4—Theoretical existence regions of Al-F complexes as a function of fluoride ion concentration.[13]

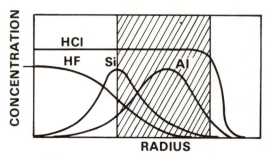

Fig. 3.5—Precipitation zone in sandstone acidizing.[13]

are believed to exist.

As a culmination of his studies, Labrid developed a model for the dynamics of equilibrium in the sandstone acidizing process. This model predicts that, near the wellbore, HF and HCl provide a solubilizing environment for the formation materials. Downstream, as shown in Fig. 3.5, the solubility of silica compounds [assumed by Labrid to be $Si(OH)_4$] is exceeded and they precipitate. This zone is delineated by the cross-hatching in Fig. 3.5. Farther downstream, with more complete dissipation of the reactant HF, fluoaluminates precipitate. Labrid and others have proposed that the permeability reduction observed experimentally when sandstone cores are first contacted with HF occurs as a result of the precipitation of silica and fluoaluminates. This reasoning is supported by the fact that the initial permeability drop in a sandstone acidizing test is more marked as the acid concentration is increased. Others, however, have attributed the permeability reduction to a largely mechanical effect. Both effects may, in fact, be active. This subject is discussed further in Chapter 9.

References

1. Ham, William E.: *Classification of Carbonate Rocks*, Memoir 1, AAPG, Tulsa, Okla. (1962).

2. Day, F. H.: *The Chemical Elements in Nature*, Reinhold Publishing Co., New York (1963) 61.

3. Goldsmith, J. R.: *Researches in Geochemistry*, John Wiley & Sons, Inc., New York (1959) 337.

4. Denbigh, K. G.: *The Principles of Chemical Equilibrium*, Cambridge University Press (1961) 268-276.

5. Lewis, G. N. and Randall, M. (revised by Pitzer, K. S. and Brewer, L.): *Thermodynamics*, McGraw-Hill Book Co., Inc., New York (1961) Chap. 15.

6. *International Critical Tables*, McGraw-Hill Book Co., Inc., New York (1933) 233.

7. Robinson, R. A. and Stokes, R. H.: *Electrolyte Solutions*, Butterworths, London (1965) 517.

8. Chatelain, J. C., Silberberg, I. H., and Schechter, R. S.: "Thermodynamic Limitations in Organic-Acid/Carbonate Systems," *Soc. Pet. Eng. J.* (Aug. 1976) 189-195.

9. Harris, F. N.: "Applications of Acetic Acid to Well Completion, Stimulation and Reconditioning," *J. Pet. Tech.* (July 1961) 637-639.

10. *Handbook of Chemistry and Physics*, 48th ed., The Chemical Rubber Co., Cleveland, Ohio (1967).

11. Perry, J. H.: *Chemical Engineers Handbook*, 3rd ed., McGraw-Hill Book Co., Inc. (1950).

12. Lenhard, P. J.: "Mud Acid — Its Theory and Application to Oil and Gas Wells," *Pet. Eng.*, Annual Issue (1943) 82-98.

13. Labrid, J. C.: "Thermodynamics and Kinetics Aspects of Argillaceous Sandstone Acidizing," *Soc. Pet. Eng. J.* (April 1975) 117-128.

Chapter 4

Reaction Kinetics

4.1 Introduction

The over-all reaction rate between acid and a formation material determines the time required for acid to react. By considering this time and the geometry within which reaction occurs, the distance acid penetrates away from the wellbore, or into a pore, and thereby the expected stimulation ratio, can be predicted.

One of the chemical reactions of primary interest in carbonate acidizing is that between hydrochloric acid and calcium carbonate.

$$2\,HCl + CaCO_3 \rightleftarrows H_2O + CO_2 + CaCl_2. \dots \dots (4.1)$$

In this reaction, two molecules of HCl in solution react with one calcium carbonate molecule at the wall of the channel through which the acid solution is flowing. Since reaction occurs between an aqueous solution and a solid, it is called a *heterogeneous reaction*. All acid reactions of interest in this monograph are heterogeneous.

The *observed reaction rate* of a component in a heterogeneous reaction is the time rate of change of the concentration of one component in the liquid. We usually refer to this as the rate of change of acid concentration. The observed rate can be controlled by one or a combination of the two steps, illustrated schematically in Fig. 4.1:

- The *rate of acid transfer* to the reactive surface by diffusion, flow-induced mixing (forced convection), mixing resulting from density gradients (free convection), or fluid loss into the formation.
- The *kinetics of the surface reaction* once acid reaches the rock surface.

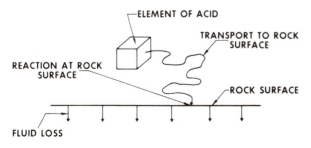

Fig. 4.1—System in which acid reaction occurs.

To understand the over-all reaction process it is necessary to develop an understanding of these mechanisms that control the *kinetics* (time dependence) of the reaction.

In this chapter, acid mass transfer and surface reaction kinetics are discussed; the important rate controlling parameters are identified; and models for the processes developed. Procedures for evaluating the kinetic parameters also are discussed.

4.2 Surface Reaction Kinetics

General Reaction Model

Heterogeneous reactions are complex, and usually are composed of several steps. In acid reactions (illustrated by Eq. 4.1) the liquid-phase reactant (HCl) must adsorb on the rock surface at a reactive site, combine with the solid, and then reaction products must be desorbed to free the surface for continued reaction. This process can be described in terms of absolute rate theory as proposed by Laidler;[1] however, this approach seldom is used because considerable experimental data are required.

A simpler approach, and the one recommended here, is to relate the rate of reaction at the rock surface to the chemical activity of components in the solution. For example, in the reaction of HCl with $CaCO_3$, if the production rate of HCl is dependent only on HCl activity, the rate is

$$r_{HCl} = -\,\xi_f a_{HCl}{}^m. \dots \dots \dots \dots \dots \dots \dots (4.2)$$

(Note: the negative sign indicates that HCl is disappearing rather than being produced by the chemical reaction.)

The exponent on the activity, m, is called the *reaction order*. If $m = 1$, the reaction is called a *first-order reaction*; if $m = 2$, the reaction is *second order*, and so on. The constant ξ_f is the *forward rate constant*, which is independent of the component activity and is dependent only on the system temperature and pressure. The *chemical activity* of a component, such as a_{HCl}, is a measure of its thermodynamic potential in a mixture. This activity is a driving force for change; when a difference in activity exists, the system will change to remove the difference. In an ideal solution, the activity of a component is equal to the component concentration; however, acid solutions are nonideal, as shown in Table 3.5.

If the reaction denoted by Eq. 4.1 is reversible, the reaction-rate expression must include the production rate of HCl by reaction between $CaCl_2$, H_2O, and CO_2. Noting that the activity for the solid phase is unity (that is, $a_{CaCO3} = 1$), this expression can be written as

$$r_{HCl} = -\xi_f a_{HCl}{}^{m_1} + \xi_r (a_{CaCl_3}{}^{m_2} \cdot a_{H_2O}{}^{m_3}$$

$$\cdot\; a_{CO_2}{}^{m_4}). \quad\ldots\ldots\ldots\ldots\ldots\ldots\ldots\ldots (4.3)$$

Here, ξ_r is the rate constant for the reverse reaction.

Kinetic models for other reactions may be constructed similarly. As shown in Table 4.1, many of the heterogeneous reactions of interest have a structure similar to Eq. 4.3.

Equilibrium in Surface Reactions

The stoichiometry of acid reactions, including the concept of equilibrium in acid reactions, was discussed in Chapter 3. At equilibrium, the net production rate of a component at the surface is zero (that is, $r_{HCl} = 0$) and, for example, Eq. 4.3 yields

$$\xi_f\; a_{HCl}{}^{m_1} = \xi_r\, a_{CaCl_2}{}^{m_2} \cdot a_{H_2O}{}^{m_3} \cdot a_{CO_2}{}^{m_4}. \;\ldots\ldots (4.4)$$

From thermodynamic arguments, it can be shown also that the equilibrium constant, K_{rxn}, is related to the reaction parameters ξ_f and ξ_r

$$K_{rxn} = \left(\frac{\xi_f}{\xi_r}\right)^z. \quad\ldots\ldots\ldots\ldots\ldots\ldots\ldots (4.5)$$

TABLE 4.1—KINETIC MODELS FOR ACIDIZATION REACTIONS
(All concentrations are expressed in gram-moles per cubic centimeter and all rates are moles produced per square centimeter per second.)

Reaction	Kinetic Model (Reference)	Parametric Value	Notes
1. Calcium carbonate-hydrochloric acid, $CaCO_3 + HCl$	$r_{HCl} = -\xi_f(\gamma_{HCl}c_{HCl})^m$ (Nierode and Williams[3])	$m = 0.2$; $\xi_f = \xi°\exp(-\Delta E/RT)$ $\xi° = 1.51 \times 10^5$ $\Delta E = 13.1$ kcal/gm-mol	γ_{HCl} is taken as the activity of HCl solutions at the HCl concentration. It is not intended to be the activity in the reacting mixture.
2. Calcium carbonate-hydrochloric acid, $CaCO_3 + HCl$	$r_{HCl} = -\xi_f' c_{HCl}{}^m$ (Lund et al.[4])	$m = 0.63$ $\xi_f' = \xi°\exp(-\Delta E/RT)$ $\xi° = 5.66 \times 10^8$ $\Delta E = 15$ kcal/gm-mol	The form of the rate expression may be more appropriate for dolomite (see Entry 3) according to the authors, but the range of temperatures studied was too small to evaluate all the parameters.
3. Dolomite-hydrochloric acid, $CaMg(CO_3)_2 + HCl$	$r_{HCl} = -\xi_f'(c_{HCl})^m$ (Lund et al.[5])	$m = RT/(1-aT)x_m$; $\xi_f' = \xi°\exp(-\Delta E/RT)$ $\Delta E = 22.5$ kcal/gm-mol $\xi° = 9.4 \times 10^{10}$ $a = 2 \times 10^{-3}\,°K^{-1}$; $x_m = 3.2$ kcal/gm-mol	Freundlich isotherm describes adsorption equilibria and therefore the surface is heterogeneous.
4. Microcline-hydrochloric and hydrofluoric mixtures $(K_{0.7t}Na_{0.18}Si_{0.94}Ca_{0.06}$ $Al_{0.06})\, AlSi_2O_8 + HCl + HF$	$r_K = \xi_f'(1+Kc_{HCl}{}^{0.8})\, c_{HF}{}^{1.13}$ (Fogler et al.[6])	$\xi_f' = \xi° \exp(-\Delta E/RT)$ $K = K° \exp(-\Delta E_1/RT)$ $\xi° = 20.5$ $K° = 27.7$ $\Delta E = +9.2$ kcal/gm-mol $\Delta E_1 = -1.2$ kcal/gm-mol	The rate is in terms of moles of potassium dissolved per square centimeter per second. Two temperature levels were reported. Additional data are required to verify activation energies.
5. Albite-hydrochloric and hydrofluoric acid mixtures $(Na_{0.72}K_{0.08}Si_{0.8}Ca_{0.2}$ $Al_{0.2})\,(AlSi_2O_8) + HCl + HF$	$r_{Na} = \xi_f'(1 + Kc_{HCl})\, c_{HF}$ (Lund et al.[6])	$\xi_f' = \xi° \exp(-\Delta E/RT)$ $K = K° \exp(-\Delta E_1/RT)$ $\xi° = 0.686$ $K° = 63.9$ $\Delta E = 7.8$ kcal/gm-mol $\Delta E_1 = -1.1$ kcal/gm-mol	The rate is in terms of the rate of Na^+ dissolution. Two temperature levels were studied and additional data are required to be certain that the temperature dependence of the rate expression is correct.
6. Vitreous silica-hydrofluoric acid, SiO_2(amorphous) + HF	$r_{HF} = -\xi_f' c_{HF}$ (Blumberg[7])	$\xi_f' = \xi° \exp(-\Delta E/RT)$ $\Delta E = 9$ kcal/gm-mol $\xi° = 12.0$ cm/sec	Measurement made using differential thermal analysis. Surface area determined by BET method.
7. Vitreous silica-hydrofluoric acid SiO_2(amorphous) + HF	$r_{HF} = -\xi_f' c_{HF}$ (Blumberg and Stavrinou[8])	$\xi_f' = 7.16 \times 10^{-6}$ cm/sec at 32.5°C	Surface area determined by BET method.
8. Vitreous silica-hydrofluoric acid SiO_2(amorphous) + HF	$r_{HF} = -\xi_f' c_{HF}$ (Blumberg and Stavrinou[8])	$\xi_f' = 5.35 \times 10^{-6}$ cm/sec at 32.5°C	Surface area determined by BET method.
9. Vitreous silica-hydrofluoric acid SiO_2(amorphous) − HF	$r_{HF} = -\xi_f' c_{HF}$ (Mowrey[9])	$\xi_f' = 8.4 \times 10^{-6}$ cm/sec at 30°C $\xi_f' = 3.4 \times 10^{-5}$ cm/sec at 44°C $\xi_f' = 4.39 \times 10^{-5}$ cm/sec at 70°C	The addition of HCl had no effect on reaction rate. Arrhenius plot was not valid.
10. Pyrex glass-hydrofluoric acid Pyrex glass contains[11] SiO_2 − 81 weight percent Na_2O − 4 weight percent A_2O_3 − 2 weight percent B_2O_3 − 13 weight percent	$r_{HF} = -\xi_f' c_{HF}$ (Glover and Guin[10])	$\xi_f' = \xi° \exp(-\Delta E/RT)$ $\Delta E = 9.75$ kcal/gm-mol $\xi° = 1.78 \times 10^2$	Area measured as external superficial area. Did not study influence of HCl concentration.

The exponents in Eq. 4.4 thereby are related to each other by Eq. 4.6.

$$\frac{m_1}{2} = \frac{m_2}{1} = \frac{m_3}{1} = \frac{m_4}{1} = Z. \quad \dots \dots \dots \quad (4.6)$$

Eq. 4.6 indicates that the reaction order divided by the stoichiometric coefficient as expressed by Eq. 4.1 must be a constant, so that the kinetic expression reduces to the proper equilibrium equation.

Effect of Temperature on Surface Kinetics

Since kinetic expressions (Eq. 4.3) are phenomenological and do not describe actual reaction steps, both the rate constants ξ_f and ξ_r and the m_j can depend on temperature. The temperature dependance of the rate coefficients usually follows the Arrhenius equation of

$$\ln \xi_f = \ln \xi_f^o - \frac{\Delta E}{RT} , \quad \dots \dots \dots \dots \quad (4.7)$$

and

$$\ln \xi_r = \ln \xi_r^o - \frac{\Delta E - \Delta H_{rxn}}{RT} . \quad \dots \dots \dots \quad (4.8)$$

Coefficients in these equations are obtained by plotting ξ_f and ξ_r vs $1/T$ and determining ξ_f^o, ξ_r^o, and ΔE from the slope and intercept of the plots. ΔH_{rxn} is the heat of reaction that can be determined from thermodynamic tables. If the reaction coefficients do not fit the Arrhenius equation, it is usually necessary to assume the m_j are also temperature dependent and fit them to the experimental data.

Effect of Pressure on Surface Kinetics

Acid reaction kinetics probably are not pressure dependent at pressures where CO_2 remains in solution (normally 600 psi or above). Since this condition usually is met in field applications, we will assume reactions are independent of pressure. At pressures below 600 psi, and when high acid concentrations are used, CO_2 may evolve from the reacting solution. Little is known about surface kinetics under these conditions.

Relationship Between Activities and Concentrations

The relationship between the activity of a component, a, and the concentration of all components in a mixture is generally a complex one. In many applications, this relationship is written

$$a = \gamma c, \quad \dots \dots \dots \dots \dots \dots \dots \quad (4.9)$$

where γ is an activity coefficient that depends on the component concentration and the concentration of other ions in solution.[2] Some reported kinetic studies incorporate the activity coefficient into the rate constants:

$$r_{HCl} = - \xi_f \, a_{HCl}^{m_1} = - \xi_f (\gamma_{HCl} c_{HCl})^{m_1}$$

$$= - \xi_f' \, c_{HCl}^{m_1}. \quad \dots \dots \dots \dots \dots \quad (4.10)$$

Measured Kinetic Parameters for Acid Reactions

A number of studies have been conducted to determine kinetic models for reaction of acids with formation minerals.

Results of these studies are summarized in Table 4.1. Note that the number of entries in this table is much smaller than the number of minerals found in reservoirs, since few accurate kinetic studies have been made. Some of these results should be regarded as preliminary. More experiments are needed to define fully the concentration and temperature dependence of some of the reactions.

4.3 Mass Transfer in Acid Solutions

General Mass Transfer Model

When the rate of a heterogeneous reaction is limited by the rate at which acid can reach the reactive surface, the reaction is said to be *mass-transfer limited*. The transfer of reactant to the surface and reaction products away from the surface can be controlled by either molecular processes (*diffusion*) or fluid flow (*convection*). If there is no flow, the rate of transfer of a component by diffusion alone in an ideal solution can be described approximately using Fick's law:

$$u_{A,Y} = -D_A \frac{\partial c_A}{\partial Y} . \quad \dots \dots \dots \dots \dots \quad (4.11)$$

This equation relates the *flux* of component A (quantity of A/unit time/unit area normal to the concentration gradient), $u_{A,Y}$, to the *concentration gradient*, $\partial c_A / \partial Y$, and an *ionic, or molecular diffusion coefficient*, D_A.*

In most acid treatments, the rate of acid transfer to the reactive surface is controlled by flow (convection) rather than ionic or molecular diffusion. Convective transfer is generated often as a result of turbulence, acid flow along a rough channel (fracture or wormhole), or by density differences between reacted and unreacted acid. It is usually impossible to predict theoretically the rate of acid transfer by convection in complicated flow geometries. Often the convective flux is described using Eq. 4.11 with an *effective diffusion coefficient, D_e,* determined experimentally for the expected geometry and flow conditions. These coefficients will have a minimum value equal to the ionic or molecular coefficient and can be several orders of magnitude larger. Models of this type are described in Chapter 6, Section 6.5, for acid fracturing and in Chapter 8, Section 8.7, for matrix acidizing.

In many instances, particularly when attempting to measure surface kinetics, the observed reaction rate will not be restricted totally by either surface kinetics or mass transfer but by a combination of both resistances. In these instances, mass transfer rates must be predicted accurately.

Diffusion of Strong Acids

In Chapter 3, we distinguished between "strong" and "weak" acids to predict the extent of their reaction with carbonates. The ionic character of the acid solution is also important in determining the diffusion rate for the acid. In ionic solutions, such as HCl, electrical forces are "long range", and positive and negative charges must move together to maintain a local charge balance.[12] This is different

---

*Acid solutions often are ionized and nonideal. This simple diffusion model alone often is not adequate to describe the transport process.[13]

TABLE 4.2—EXPERIMENTAL DIFFUSION COEFFICIENTS[3]
Concentration gm-mol/liter

| Experiment | Solution 1 (Surface) | | | Solution 2 (Boundary Layer) | | | Experimental D_{H^+} ($\times 10^5$ sq cm/sec) |
	H^+	Ca^{++}	Cl^-	H^+	Ca^{++}	Cl^-	
1	0.80	1.70	4.20	1.48	0.86	3.20	0.98
2	1.07	1.02	3.11	1.48	0.58	2.64	1.25
3	3.82	0.29	4.40	4.34	0.25	4.84	0.73
4	3.60	0.40	4.40	4.34	0.30	4.94	1.05
5	3.08	0.70	4.48	4.34	0.46	5.26	1.20
6	7.08	2.63	12.34	8.08	1.29	10.66	0.23
7	4.04	1.00	6.04	8.08	1.00	10.08	0.18
8	6.06	1.50	9.06	8.08	1.00	10.08	0.07

from diffusion in nonionized or weakly ionized solutions, where the rate of diffusion of a given species is essentially independent of other components in the solution.[13] Sherwood and Wei,[14] Gilliland *et al.*,[15] Vinograd and McBain,[16] and others[17, 18] have studied the problems of diffusion in complex systems and present techniques for predicting ionic diffusion rates.

To model accurately the over-all acid reaction, it is necessary to predict the rate of movement of positively charged hydrogen ions to the surface while positively charged calcium ions are moving away from the surface. Negatively charged chloride ions will be distributed to maintain local electrical neutrality. The range of hydrogen-ion diffusion coefficients possible in a reacting system is illustrated in measurements made by Nierode and Williams,[3] who studied the counter diffusion of two partially reacted hydrochloric acid solutions (summarized in Table 4.2). These solutions were selected to represent concentrations at the reactive surface and a short distance from the surface. Significant differences in hydrogen-ion diffusion rates were found as

total hydrogen-ion content was increased, and at a constant hydrogen-ion concentration, when the calcium-ion content of each solution was changed.

Because of the difficulty of theoretically predicting the diffusional behavior of concentrated hydrochloric acid solutions, Nierode and Williams determined effective hydrogen-ion diffusion rates experimentally. In the absence of precise experimental data, Roberts and Guin[19] used an effective diffusion coefficient model that gives the results plotted in Fig. 4.2. We will use Roberts and Guin's diffusion coefficients for calculations in examples to follow, even though they are often not accurate. More accurate predictions can be made using techniques described by Miller,[17, 18] although these are beyond the scope of this monograph.

Diffusion of Weak Acids

Weak acids are ionized only slightly in aqueous solution; therefore, their diffusion rates can be represented as a sum of the ion flux and the flux of the neutral molecule. Usually, the diffusion of neutral species will dominate because the molecular concentration is much higher than the ionic concentration. The diffusivity of the molecules can be determined using techniques described by Harned and Owens.[20] As with strong acids, the recommended procedure in most applications is to use an experimentally determined effective diffusivity. Diffusivities of the more common weak acids are shown in Fig. 4.3.

Convective Mass Transfer

Laminar Flow — The rate of mass transfer in a laminar flow is determined by both diffusion and flow. The rate of acid transfer to a reactive surface, the mass flux $u_{A,Y}$, is given by Fick's law (Eq. 4.11) with a term added to account for acid transfer to the surface because of flow normal to the surface ($c_A v_N$), where v_N is the fluid velocity normal to the surface.

$$u_{A,Y} = -D_A \frac{dc_A}{dY} + c_A v_N. \quad\quad (4.12)$$

Turbulent Flow — In turbulent flow, the velocity and concentration at a point fluctuate in a seemingly random fashion around their mean values. For a turbulent system, Fick's law can be written in terms of the time-averaged components (indicated by < >):

$$\langle u_{A,Y} \rangle = -D_A \frac{d\langle c_A \rangle}{dY} + \langle c_A v_N \rangle, \quad\quad (4.13)$$

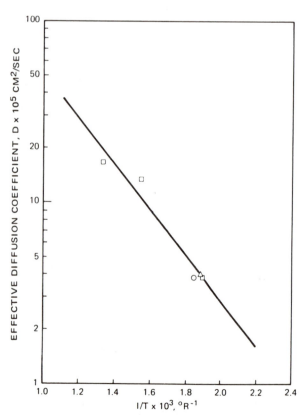

Fig. 4.2—Effective diffusivity for 1.4 N HCl.[19]

Or, expanding in terms of the fluctuating components and

defining an "effective diffusivity" or "eddy diffusivity," D_e, we obtain

$$<u_{A,Y}> = -D_e \frac{d<c_A>}{dY} + <c_A><v_N>. \quad \ldots \ldots (4.14)$$

The effective diffusivity, D_e, is a complex function best described by an expression developed by Notter and Sleicher.[24] Since their model is more complex than normally is required for acidizing applications, a simpler model will be presented. In this simple model, the flux to the surface, $u_{A,Y}$, is related to the difference between the average concentration in the flowing acid, $<c_A>$, and the average wall concentration, $<c_A(w)>$, by a mass transfer coefficient, K_g.

$$u_{A,Y} = K_g \left[<c_A> - <c_A(w)> \right] + <c_A><v_N>.$$

$$\ldots \ldots \ldots \ldots \ldots \ldots \ldots \ldots \ldots \ldots \ldots \ldots (4.15)$$

The effective mass transfer coefficient, K_g, generally is determined experimentally and correlated as a function of the Reynolds and Schmidt numbers. Results of studies to determine K_g for a fracture system are shown in Chapter 6, Section 6.5.

4.4 Models for Heterogeneous Reactions in Laminar Flow Systems

To determine the surface reaction rate from flow experiments, it is necessary to predict the extent to which mass transfer controls the reaction. Surface kinetics can be determined from tests in (1) a flow reactor in which the fluid velocity distribution is well defined and diffusion coefficients are known or can be measured independently, or (2) tests where the fluid is so highly agitated that the rate of acid transfer to the reactive wall is faster than the reaction rate at the wall. Examples of these reaction systems are illustrated schematically by parallel plate, tubular, and rotating disk reactors in Fig. 4.4. The so-called "static reaction-rate test,"[25] in which a cube of rock is added to an unstirred acid, is not suitable for kinetic determinations because the rate of acid transfer to the rock is controlled by free convection of an unknown magnitude.[31] See Chapter 6, Section 6.5, for a further discussion of this test.

In the dynamic tests just described, one or more of the walls of the system will be reactive. During the test the average reactant and product concentration will be monitored. In a flow test the effluent concentration will be observed, whereas in the rotating disk test the average concentration in the reactor will be measured. The over-all reaction rate is obtained as follows.

$$r_{\text{over-all}} = \frac{\Delta c q}{A} \quad \text{(flow system)},$$

$$r_{\text{over-all}} = \frac{\Delta c V_R}{\Delta t A} \quad \text{(rotating disk)},$$

where q is the volumetric flow rate, A is the reactive surface area, V_R is the volume of the reactor, Δc is the change in acid concentration, and Δt is the time over which the concentration change is observed.

The separation of mass transfer and surface kinetics is accomplished by developing a model for the reaction process.

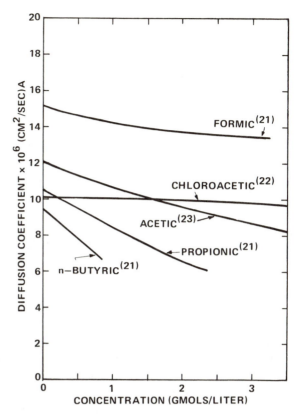

Fig. 4.3—Diffusion coefficients of weakly ionized acids at 25°C.

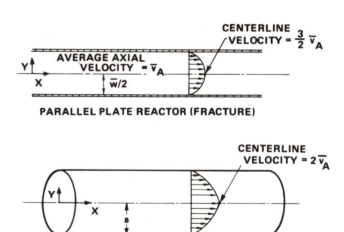

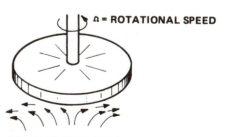

Fig. 4.4—Reaction systems suitable for studies of heterogeneous reactions.

General Procedure for Modeling the Reaction Process

To predict, or model, the rate of acid reaction in a laminar flow system, it is necessary to solve the material balance equation (Eq. 4.16) subject to the boundary conditions given as Eq. 4.17.

Material Balance

$$\frac{\partial c_A}{\partial t} = -\frac{\partial}{\partial Y}(u_{A,Y}) - \frac{\partial}{\partial X}(u_{A,X}) \quad \text{(parallel plate)}.$$

$$\dotfill (4.16)$$

Boundary Conditions

$$u_A = -r_A \quad \text{at the wall.} \dotfill (4.17)$$

To solve Eq. 4.16, it is first necessary to obtain the fluid velocity by solving the appropriate equations of motion. Many useful solutions of these equations are given by Bird et al.[26]

Parallel Plate Reactor

To illustrate procedures used to model the reaction process in flow systems, an analysis for a parallel plate reactor will be presented in detail. Models for other flow geometries can be developed in a similar fashion. The parallel plate system has been used by Barron et al.,[27] Nierode and Williams,[3] and Roberts and Guin.[28] In this analysis, it will be assumed that the mass transfer and chemical reaction have no effect on the fluid velocity, and the equations of motion therefore can be solved separately. To realize this condition in practice, the average fluid velocity down the tube should be large relative to the velocity normal to the surface created by the chemical reaction. Furthermore, variations in the fluid viscosity and gap width in the reactor are neglected. Therefore, the acid concentration change between the inlet and exit must be limited.

For the geometry shown in Fig. 4.4, the steady-state mass balance (Eq. 4.16) reduces to

$$\frac{3}{8}\frac{\bar{v}_A}{L}(1-y^2)\frac{\partial c}{\partial x} = \frac{D}{\bar{w}^2}\frac{\partial^2 c}{\partial y^2}, \dotfill (4.18)$$

where $x = X/L$ and $y = 2Y/\bar{w}$. The following assumptions were made to reduce the equation to this form.

1. The velocity profile is assumed to be laminar and not influenced by the reaction taking place at the solid surface. Actually, this assumption may be in error, depending on the relationship between the densities of the reaction products relative to the densities of the reactants (see Chapter 6, Section 6.5). In studying the reaction of HCl with $CaCO_3$, Barron et al.[27] and Nierode and Williams[3] have noted the existence of natural convection caused by the reaction product being heavier than the reactants. This would manifest itself, for example, in the mass transfer from a horizontal upper surface being greater than that at a lower surface, where there would be no induced convection patterns. Chang et al.[29] have recently studied this problem.

2. Flow is developed fully and is laminar throughout the entire test section. The reactor system would, therefore, have to include an entry region that is constructed of inert walls, permitting the velocity profile to develop before reac-

tion starts. This entry region must be designed carefully if experimental results are used to evaluate the surface reaction mechanism. Schlicting[30] has given the following expression for the required entry-region length.

$$\frac{\ell_{\text{entry}}}{\bar{w}} = 0.02 N_{\text{Re}}. \dotfill (4.19)$$

The boundary conditions imposed on the integration of Eq. 4.18 follow from Eq. 4.17 and may be written as

$$\frac{2D}{\bar{w}}\frac{\partial c}{\partial y} = r \quad \text{at } y = 1, \dotfill (4.20)$$

$$\frac{2D}{\bar{w}}\frac{\partial c}{\partial y} = -r \quad \text{at } y = -1. \dotfill (4.21)$$

For a heterogeneous, irreversible reaction involving a single reactant, Eq. 4.2 applies and the boundary conditions (Eqs. 4.20 and 4.21) reduce to

$$\frac{\partial c}{\partial y} = -\frac{\xi_f \bar{w}}{2D}a^m = -\frac{\xi_f' \bar{w}}{2D}c^m \quad \text{at } y = 1, \dotfill (4.22)$$

$$\frac{\partial c}{\partial y} = \frac{\xi_f \bar{w}}{2D}a^m = \frac{\xi_f' \bar{w}}{2D}c^m \quad \text{at } y = -1. \dotfill (4.23)$$

Eq. 4.10 was used to replace the activities with concentrations. The equations may be written as

$$\frac{\partial c^*}{\partial y} = -P_f c^{*^m} \quad \text{at } y = 1, \dotfill (4.24)$$

$$\frac{\partial c^*}{\partial y} = P_f c^{*^m} \quad \text{at } y = -1, \dotfill (4.25)$$

where the reaction parameter, P_f, is defined as

$$P_f = \frac{\bar{w}\,\xi_f'\,c_o^{m-1}}{2D}, \dotfill (4.26)$$

and $c^* = c/c_o$. For a first-order reaction ($m = 1$), $P_f = \bar{w}\,\xi_f'/2D$. c_o is the inlet concentration.

The reaction parameter, P_f, has an important physical meaning. If P_f is large, the reaction rate is very fast, relative to molecular diffusion. On the other hand, a small value of P_f indicates that the reaction rate is controlling. The subscript "f" indicates the reaction rate parameter for flow and reaction between parallel flat plates.

Eq. 4.18 has been solved with a number of different reaction mechanisms imposed as boundary conditions.[3,28,31,32] Methods of solution include the representation of concentration profiles as an infinite sum of orthogonal functions and the use of finite-difference techniques. If a flexible computer program that can be used in conjunction with a number of acid reaction mechanisms is required, the finite-difference technique seems preferable. Such a program can be written to accommodate a variety of boundary conditions. An orthogonal expansion works best when the surface reaction rate is first order in the reactant concentration, since the problem is linear in this case. Nonlinear reaction rates can be solved in terms of an integral equation, but the calculations required to obtain numerical results are tedious.[31,32]

Solutions of Eq. 4.18 are best presented graphically in terms of the dimensionless reactor length L_{fD}, defined by Eq. 4.27, the dimensionless reaction rate parameter, P_f, and a dimensionless over-all conversion, R_f^*, defined by Eq. 4.28

$$L_{fD} = \frac{8DL}{3\bar{v}_A \bar{w}^2} = \text{the dimensionless reactor length,}$$

$$\dots\dots\dots\dots\dots\dots\dots\dots \quad (4.27)$$

$$R_f^* = 1 - \frac{c}{c_o} \quad \dots\dots\dots\dots\dots\dots\dots \quad (4.28)$$

Resulting plots for first-order, $m = 1$, and fractional-order reactions are given in Figs. 4.5 through 4.8. These plots can be used to estimate acid penetration in a laboratory model, or along a fracture for the limiting case of zero fluid loss. A model including fluid loss is developed in Chapter 6 to simulate the fracturing process accurately.

Parallel Plate Reactors With One Reactive Surface

The reaction products resulting from the HCl-carbonate reaction are more dense than the reactants; therefore, natural convection may alter the velocity profile and negate the validity of the analysis given in the preceding section. In their studies, Nierode and Williams[3] circumvented this problem by using a parallel plate reactor oriented horizontally with an inert upper plate and a reactive lower plate. There are no convection currents associated with this reaction system, and it can be modeled accurately with Eq. 4.18. Of course, the appropriate boundary conditions must be imposed. Nierode and Williams presented an eigenvalue solution to this problem. The reader interested in the details of their analysis should consult their paper.

Circular Tube Laminar Flow Reactor

A solution similar to that presented for flow and reaction between parallel plates also can be developed for the tube flow reactor shown in Fig. 4.4.

The material balance equation can be solved in a fashion similar to that described for flow between parallel plates. To simplify the solution, the dimensionless reaction rate, P_c, the dimensionless tube length, L_{cD}, and the over-all conversion, R_c^*, are defined as

$$P_c = \frac{a\xi_f' c_o^{m-1}}{D} , \quad \dots\dots\dots\dots\dots \quad (4.29)$$

$$L_{cD} = \frac{DL}{2\bar{v}_A a^2} , \quad \dots\dots\dots\dots\dots \quad (4.30)$$

$$R_c^* = 1 - \frac{c}{c_o} . \quad \dots\dots\dots\dots\dots \quad (4.31)$$

Procedures for solving this problem are available.[31] Results for a first-order reaction are presented in Fig. 4.9. This model is applicable to laboratory test data or to wormholes in the limit of zero fluid loss from the wormhole.

Example — Determination of Kinetic Parameters From a Parallel Plate Reactor

To illustrate the use of Figs. 4.5 through 4.9, an example showing the determination of reaction constants will be considered. Assume that a 15-weight-percent HCl solution is made to flow at an average velocity of 0.2 in./sec between parallel limestone slabs that are spaced 0.2 in. apart. The

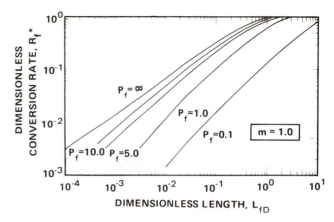

Fig. 4.5—First-order surface reactions on parallel plates.[31]

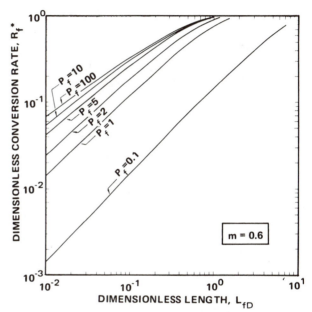

Fig. 4.6—Fractional-order surface chemical reactions on parallel plates.[32]

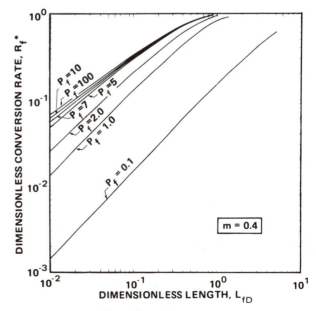

Fig. 4.7—Fractional-order surface chemical reactions in parallel plate laminar flow reactors.[32]

plates are 3 ft long. If the outlet HCl concentration is 13 weight percent, what is the forward reaction rate constant, ξ_f'? The reaction is assumed to be first order to simplify this example.

For this case,

$\bar{w} = 0.2$ in. $= 0.508$ cm

$\bar{v}_A = 0.2$ in./sec $= 0.508$ cm/sec

$L = 3$ ft $= 91.44$ cm

$D = 5.1 \times 10^{-5}$ sq cm/sec (from Fig. 4.2 at 100°F).

First, calculating the conversion R_f* and the dimensionless reactor length L_{fD} gives

$$R_f* = 1 - \frac{13 \text{ weight percent}}{15 \text{ weight percent}} = 0.133$$

$$L_{fD} = \frac{8DL}{3\bar{v}_A\bar{w}^2} = \frac{(8)(5.1 \times 10^{-5} \text{ sq cm/sec})(91.44 \text{ cm})}{(3)(0.508 \text{ cm/sec})(0.508 \text{ cm})^2}$$

$$= 0.098.$$

Locating this point on Fig. 4.5, it is seen to lie between the $P_f = 5.0$ and $P_f = 1.0$ curves at approximately $P_f = 3.0$.

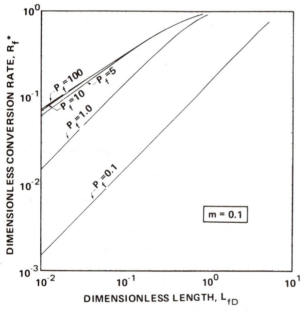

Fig. 4.8—Fractional-order surface chemical reactions in parallel plate laminar flow reactors.[32]

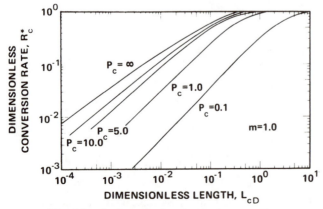

Fig. 4.9—Surface reactions in circular pores.[31]

Solving Eq. 4.26 for ξ_f' gives

$$\xi_f' = \frac{2DP_f}{\bar{w}} = \frac{(2)(5.1 \times 10^{-5} \text{ sq cm/sec})(3.0)}{0.508 \text{ cm}}$$

$$= 6.02 \times 10^{-4} \text{ cm/sec.}$$

The reaction-rate expression valid for this particular experiment, assuming that the rate is first order in acid concentration, is therefore

$$r_{HCl} = -(6.02 \times 10^{-4} \text{ cm/sec}) c_{HCl}.$$

Note that with only a single experiment, or a few experiments, the data often can be fitted with several kinetic models. For this reason, a rate expression should not be accepted without performing a number of measurements and obtaining the best fit of all data.

Example — Prediction of Acid Penetration Distance Along a Circular Tube

In the previous example for reaction between parallel plates, the reaction-rate expression was found to be

$$r_{HCl} = -(6.02 \times 10^{-4} \text{ cm/sec}) c_{HCl}.$$

Calculate the distance that acid will penetrate down a circular hole 0.1 cm in radius before it is 95-percent reacted, assuming the kinetic model given above. The average velocity is 0.1 cm/sec.

The reaction was assumed to be first order, therefore, Fig. 4.9 applies. To use this figure, we need to calculate P_c and R_c*.

$$P_c = \frac{(a)(\xi_f')}{D} = \frac{(0.1 \text{ cm})(6.02 \times 10^{-4} \text{ cm/sec})}{(5.1 \times 10^{-5} \text{ sq cm/sec})}$$

$$= 1.18.$$

If the acid is to be 95-percent reacted,

$$R_c* = 1 - 0.05 = 0.95.$$

From Fig. 4.9,

$$L_{cD}* = 1.2.$$

Solving for L,

$$L = \frac{2\bar{v}_A a^2 L_{cD}*}{D} = \frac{(2)(0.1 \text{ cm/sec})(0.1 \text{ cm})^2 (1.2)}{5.1 \times 10^{-5} \text{ sq cm/sec}}$$

$$= 47 \text{ cm.}$$

The Rotating Disk System

As previously noted, a preferred procedure for determining surface kinetics is to use a system where the rate of acid transfer to the reactive surface greatly exceeds the reaction rate at the surface. One system for which these conditions can often be satisfied is the rotating disk system shown in Fig. 4.4. This system consists of a large disk that is immersed in the fluid and rotated around its center. Because of the rotation, there is a centrifugal effect that will cause the fluid to flow in a radial direction along the disk and reduce the thickness of the stagnant fluid layer near the surface (often called the diffusion boundary layer). According to Levich,[33] each point on the disk surface is then equally accessible.

In designing a rotating disk reactor system, the disk radius must be much larger than the boundary-layer thickness (often about 0.01 cm) and flow in the system must be laminar. In theory, the system is designed so that the disk spins in a infinite volume; however, Gregory and Riddiford[34] demonstrated that, in a liquid system, if the vessel diameter is at least twice the disk diameter, the observed rate is independent of the vessel diameter. Boomer et al.[35] have reported the details of a system suitable for studying acid reaction kinetics.

Since each point on the surface is equally accessible, the following equality between reaction rate and mass transfer rate must apply.

$$\frac{D_j}{\delta}\left(c_j^{(\infty)} - c_j^s\right) = -r_j. \quad \dots \dots \dots \dots \dots \dots \quad (4.34)$$

c_j^∞ is the concentration of the reactive species in the bulk fluid and δ is the diffusion boundary-layer thickness. At small, angular velocities δ is large, the reaction is diffusion controlled, and the difference between the bulk and surface concentrations is large. As rotational speed is increased, the boundary-layer thickness decreases, thereby decreasing the mass transfer resistance.

A typical plot of reaction rate vs rotational speed is shown schematically in Fig. 4.10. When the reaction rate no longer increases with an increase in rotational speed, the observed reaction rate is caused entirely by the surface kinetics. Under these conditions, the surface kinetics are obtained directly, and diffusion rates do not have to be calculated as was done for tubular and parallel plate reactors. This system was used by several authors to measure the surface reaction kinetics of many acid-mineral systems[4-6,36] (see also Table 4.1).

References

1. Laidler, K. J.: Chemical Kinetics, 2nd ed., McGraw-Hill Book Co., Inc., New York (1965) 286-296.

2. Wall, F. T.: Chemical Thermodynamics, W. H. Freeman and Co., San Francisco (1965) 372-375, 401-410.

3. Nierode, D. E. and Williams, B. B.: "Characteristics of Acid Reactions in Limestone Formations," Soc. Pet. Eng. J. (Dec. 1971) 406-418.

4. Lund, K., Fogler, H. S., McCune, C. C., and Ault, J. W.: "Acidization II: The Dissolution of Calcite in Hydrochloric Acid," Chem. Eng. Sci. (1975) 30, 825-835.

5. Lund, K., Fogler, H. S., and McCune, C. C.: "Acidization I: The Dissolution of Dolomite in Hydrochloric Acid," Chem. Eng. Sci. (1973) 28, 681-700.

6. Fogler, H. S., Lund, K., McCune, C. C., and Ault, J. W.: "Dissolution of Selected Minerals in Mud Acid," paper 52c, presented at the AIChE 74th National Meeting, New Orleans, March 11-15, 1973.

7. Blumberg, A. A.: "Differential Thermal Analysis and Heterogeneous Kinetics: The Reaction of Vitreous Silica with Hydrofluoric Acid," J. Phys. Chem. (1959) 63, 1129.

8. Blumberg, A. A. and Stavrinou, S. C.: "Tabulated Functions for Heterogenous Reaction Rates: The Attack of Vitreous Silica by Hydrofluoric Acid," J. Phys. Chem. (1960) 64, 1438.

9. Mowrey, S. L.: "The Theory of Matrix Acidization and the Kinetics of Quartz-Hydrogen Fluoride Acid Reactions," MS thesis, U. of Texas, Austin (1974). Also available as Report No. UT 73-4, Texas Petroleum Research Committee, Austin (1974).

10. Glover, M. C. and Guin, J. A.: "Permeability Changes in a Dissolving Porous Medium," AIChE Journ. (1973) 19, 1190.

11. Bolz, R. E. and Tuve, G. L. (ed.): Handbook of Tables for Applied Engineering Science, Chemical Rubber Co., Cleveland, Ohio (1970).

12. Resibois, P. M. V.: Electrolyte Theory, Harper and Row, New York (1968) Ch. 3.

13. deGroot, S. R. and Mazur, P.: Non-Equilibrium Thermodynamics, North-Holland Publishing Co., Amsterdam (1962) 367-375.

14. Sherwood, T. K. and Wei, J. C.: "Ion Diffusion in Mass Transfer Between Phases," AIChE Journ. (1955) 4, 1.

15. Gilliland, E. R., Baddour, R. F., and Goldstein, D. J.: "Counter Diffusion of Ions in Water," Cdn. J. Chem. Eng. (1957) 37, 10.

16. Vinograd, J. R. and McBain, J. W.: "Diffusion of Electrolytes," J. Am. Chem. Soc. (1941) 63, 2009.

17. Miller, D. G.: "Application of Irreversible Thermodynamics Electrolyte Solutions I: Determination of Ionic Transport Coefficients $_{ij}$ for Isothermal Vector Transport Processes in Binary Electrolyte Systems," J. Phys. Chem. (1966) 70, 2639.

18. Miller, D. G.: "Application of Irreversible Thermodyanamics to Electrolyte Solutions II: Ionic Coefficients ℓ_{ij} for Isothermal Vector Transport Processes in Ternary Systems," J. Phys. Chem. (1967) 71, 66.

19. Roberts, L. D. and Guin, J. A.: "A New Method for Predicting Acid Penetration Distance," Soc. Pet. Eng. J. (Aug. 1975) 277-286.

20. Harned, H. S. and Owens, B.: Physical Chemistry of Electrolytic Solutions, 3rd ed., Reinhold Publishing Co., New York (1958).

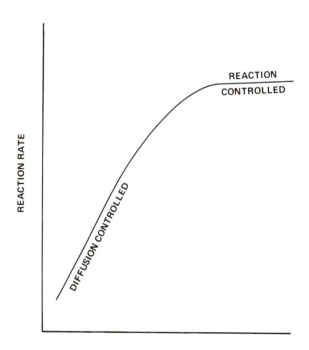

Fig. 4.10—Typical reaction rate measurement as a function of rotation speed of reactive disk.

21. Dunn, L. A. and Stokes, R. H.: "Diffusion Coefficients of Monocarboxylic Acids at 25°C," *Aust. J. Chem.* (1965) **18,** 285.

22. Garland, C. W., Tong, S., and Stockmayer, W. H.: "Diffusion of Chloroacetic Acid in Water," *J. Phys. Chem.* (1965) **69,** 2469.

23. Vitagliano, V.: "Diffusion in Aqueous Acetic Acid Solutions," *J. Am. Chem. Soc.* (1955-56) **78,** 4538.

24. Notter, R. H. and Sleicher, C. A.: "The Eddy Diffusivity in the Turbulent Boundary Layer Near a Wall," *Chem. Eng. Sci.* (1972) 2073-2093.

25. Hendrickson, A. R., Rosene, R. B., and Wieland, D. R.: "Acid Reaction Parameters and Reservoir Characteristics Used in the Design of Acid.Treatments," paper presented at the ACS 137th Meeting, Cleveland, Ohio, April 5-14, 1960.

26. Bird, R. B., Stewart, W. E., and Lightfoot, E. N.: *Transport Phenomena,* John Wiley & Sons, Inc., New York (1960).

27. Barron, A. N., Hendrickson, A. R., and Wieland, D. R.: "The Effect of Flow on Acid Reactivity in a Carbonate Fracture," *J. Pet. Tech.* (April 1962) 409-415; *Trans.,* AIME, **225.**

28. Roberts, L. D. and Guin, J. A.: "The Effect of Surface Kinetics in Fracture Acidizing," *Soc. Pet. Eng. J.* (Aug. 1974) 385-395; *Trans.,* AIME, **257.**

29. Chang, C. Y., Guin, J. A., and Roberts, L. D.: "Surface Reaction with Combined Forced and Free Convection," *AIChE Journ.* (1976) **22,** 252.

30. Schlichting, H.: *Boundary Layer Theory* (translated by J. Kestin), Pergamon Press, New York (1955) 146-149.

31. Williams, B. B., Gidley, J. L., Guin, J. A., and Schechter, R. S.: "Characterization of Liquid-Solid Reactions: The Hydrochloric Acid-Calcium Carbonate Reaction," *Ind. and Eng. Chem. Fund.* (1970) **9,** 589-596.

32. Guin, J. A.: "Chemically Induced Changes in Porous Media," PhD dissertation, U. of Texas, Austin (Jan. 1970). Also Report No. UT 69-2, Texas Petroleum Research Committee, Austin (1969).

33. Levich, V. G.: *Physicochemical Hydrodynamics,* Prentice-Hall, Inc., Englewood Cliffs, N.J. (1962) 60-75.

34. Gregory, D. P. and Riddiford, A. C.: "Transport to the Surface of a Rotating Disc," *J. Chem. Soc.* (1956) 3756.

35. Boomer, D. R., McCune, C. C., and Fogler, H. S.: "Rotating Disk Apparatus for Studies in Corrosive Liquid Environments," *Review Scientific Instruments* (1972) **43,** 225.

36. Fogler, H. S., Lund, K., McCune, C. C., and Ault, J. W.: "Kinetic Rate Expressions for Reactions of Selected Minerals with HCl and HF Mixtures," paper SPE 4348 presented at the SPE-AIME Oilfield Chemistry Symposium, Denver, May 24-25, 1973.

Chapter 5

Acid Fracturing Fundamentals

5.1 Introduction

As discussed in Chapter 2, Section 2.3, the stimulation ratio resulting from an acid fracturing treatment is controlled by two characteristics of the fracture formed by acid reaction: the length of the conductive fracture that remains open after the treatment, and the ratio of the conductivity of this fracture to the formation permeability. The acidized fracture length and fracture conductivity are controlled largely by the treatment design and formation strength. The relationship among the stimulation ratio, these variables, and formation characteristics was illustrated in Fig. 2.5.

To design acid fracturing treatments effectively, it is necessary to understand how various parameters alter the treatment. This chapter presents these relationships in qualitative terms. The sequence of presentation corresponds to steps in a treatment. The effect of fluid and formation characteristics on the geometry of a fracture created by the acid, or a pad fluid before acid injection, is described first. This is followed by a discussion of variables that control the distance reactive acid can move along a fracture (called the acid penetration distance) and the conductivity of the fracture created as a result of acid reaction.

Procedures for mathematically modeling each step in the acid fracturing process are detailed in Chapter 6, and a design procedure using these models is discussed in Chapter 7.

5.2 Fracture Geometry

In an acid fracturing treatment, acid, or a fluid used as a pad before the acid, is injected down the well casing or tubing at rates higher than the reservoir will accept. This produces a buildup in wellbore pressure until it exceeds the compressive earth stresses and tensile rock strength. At this pressure the formation fails, allowing a crack (fracture) to be formed. The fracture so formed is propagated by continued fluid injection.

A fracture propagates in the orientation that requires the least amount of work. Since subsurface rocks are under compressive stresses because of the weight of overburden and forces generated by tectonic activity, a fracture will orient itself so that it grows perpendicular to the axis of the smallest principle compressive stress. Normally, the verti-

cal stress will be larger than either horizontal stress, so the fracture will be oriented vertically, as shown in Fig. 5.1. In very shallow wells (normally, less than 2,000 ft), horizontal stresses sometimes can exceed the vertical stress, in which case the fracture can propagate in a horizontal plane. In this monograph the fracture is assumed to be vertical.

The geometry of the fracture formed by fluid injection is controlled by the elastic properties of the formation rock, the force generated by the pressure increase in the fracture because of the frictional resistance to fluid flow along the fracture (called the fracture propagation pressure), and the rate of fluid flow into the formation relative to the rate of fluid injection. The ratio of the fracture width measured at the wellbore to the fracture length is proportional to fluid, formation, and fracture properties. One theoretical result relating width, length, and formation properties is given in Eq. 5.1.[1]

$$\frac{w_w}{L} \approx \left(\frac{\mu i}{EhL^2}\right)^{0.25} \quad \dots\dots\dots\dots\dots\dots\dots (5.1)$$

This relationship shows that the fracture width will increase

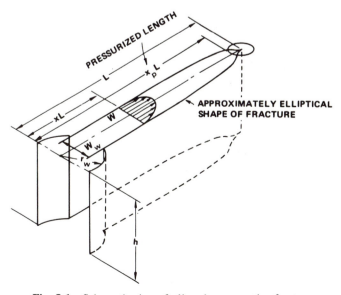

Fig. 5.1—Schematic view of a linearly propagating fracture.

if the fracture length, *L*, the fluid viscosity, μ, or the injection rate per unit of fracture height, *i/h*, is increased, and will decrease as the Young's modulus, *E*, of the formation increases. Many carbonate formations have large Young's moduli and therefore do not easily allow wide fractures to be formed. The fracture width can be maximized in such formations by (1) using a pad fluid with a high viscosity, (2) injecting fluid at a high rate, (3) injecting a large volume of fluid, and/or (4) reducing the rate of fluid loss to the formation by adding fluid-loss additives to give a larger fracture volume after a given volume of fluid injection. Models for predicting fracture geometry are presented in Chapter 6, Section 6.2.

In an acid fracturing treatment, two procedures are commonly used. Acid alone is injected, or a fluid (called a pad) that will create a long, wide fracture is injected and followed by acid. When acid is used without a pad fluid, the dynamic fracture often will be short and narrow because the rate of fluid loss for acid alone is generally high and, as a result of its low viscosity, the fracture propagation pressure will be low. A long, wide fracture usually can be created by selecting a high-viscosity pad fluid that can control the rate of fluid loss to the formation. After acid injection is initiated, the geometry will be altered, depending on the acid fluid-loss characteristics, as discussed in the next section.

5.3 Acid Penetration Distance

The distance that reactive acid moves along a fracture during the treatment (called the acid penetration distance) is one of the variables that will determine the success or failure of the treatment. This distance is controlled by the acid fluid-loss characteristics, the rate of acid reaction with the formation rock, and the acid flow rate along the fracture.

As discussed in Chapters 4 and 6, the over-all reaction rate in most carbonates is controlled largely by the rate of acid transfer to the fracture walls, not by the surface reaction kinetics. (The reaction system is shown schematically in Fig. 5.2.) The rate of acid transfer to the fracture wall in turn is controlled by the distance acid must move to reach the wall (that is, the fracture width), the rate of acid flow to the wall because of fluid loss to the formation, and any mixing that may occur in the fracture. As shown schematically in Fig. 5.3, at low flow rates, flow is laminar and acid transfer to the fracture wall will be influenced by secondary flow induced by density changes resulting from reaction. At higher flow velocities, the flow becomes turbulent and the rate of acid transfer will increase as injection rate increases.[2,3]

Effect of Acid Fluid-Loss Rate

When acid enters a fracture it will react with the fracture walls and can eliminate the filter cake created by fluid-loss additives used in the pad fluid.[2] Once this occurs, the fracture geometry will be controlled primarily by the acid fluid-loss characteristics. The mechanism by which acid bypasses the fluid-loss additive is shown in Fig. 5.4. This photograph shows "wormholes" formed in the face of a limestone fracture wall as a result of acid leak-off perpendicular to the face.

The change in fracture geometry after acid injection into a fracture created by a viscous pad fluid cannot be predicted accurately. It is safe to conclude, however, that if an effective fluid-loss additive is not included in the acid, the rate of fluid loss after acid injection begins will exceed that when only pad was injected, and eventually the fracture will begin to close because of this increased rate of fluid loss.

The expected general behavior is depicted in Fig. 5.5. Case 1 in this schematic is expected if the acid has a somewhat higher fluid-loss rate than the pad fluid, and may represent what would occur in a very low-permeability for-

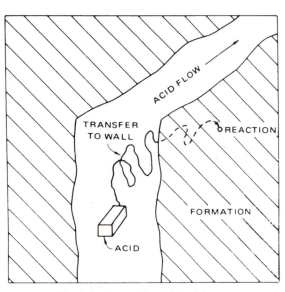

Fig. 5.2—System in which acid reacts.

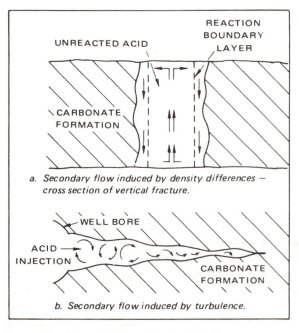

a. Secondary flow induced by density differences — cross section of vertical fracture.

b. Secondary flow induced by turbulence.

Fig. 5.3—Acid flow behavior in the fracture.

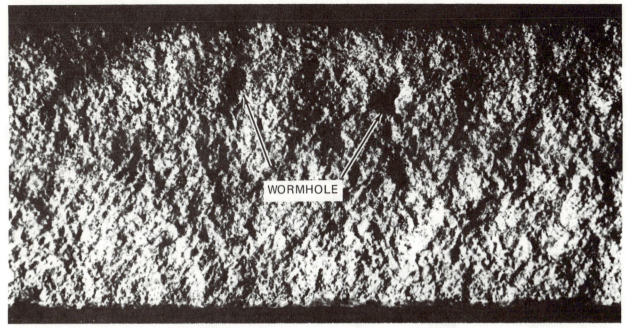

Fig. 5.4—Surface of a fracture after acid reaction.

mation. In this example, the fracture will continue to grow for a time, but then will be reduced in length as acid injection is continued. The fracture geometry ultimately will approach the geometry expected if the pad fluid had not been used. Case 2 would be expected if the rate of acid fluid loss is much higher than that for the pad fluid and generally represents what is expected in a high-permeability formation. Again, the fracture geometry ultimately will approach that expected if acid alone is used. If the acid contains an additive that gives it fluid-loss chracteristics comparable with the pad fluid, it is usually safe to assume that the fracture will continue to increase in volume and length during acid injection.

A reduction in the open fracture length after acid injection begins will be accompanied by a reduced fracture width, as indicated by Eq. 5.1. The effect of a change in fracture width on the acid penetration distance is discussed later in this section.

It is clear that the use of an effective fluid-loss additive in the acid is central to maximizing the acid penetration distance. Fluid-loss control of acid in carbonates is generally much more difficult than when fracturing a sandstone with an inert fluid because the acid continually dissolves the rock matrix that supports the fluid-loss additive. In addition, many carbonates are naturally fractured or vugular and therefore have flow channels that are much harder to plug with an additive. Fig. 5.6 shows the inlet and outlet faces of three cores used for fluid-loss testing of acids.[5] Cores 1 and 2 were used in a fluid-loss test at 140°F with 1,000-psi differential pressure, and Core 3 is the result of a similar test at 200°F. A fluid-loss additive was used in the tests on Cores 2 and 3. These experiments show that an effective fluid-loss additive can cause acid reaction to be more uniform, produce several wormholes instead of just one, as shown in Core 1, and thereby greatly restrict the rate of fluid loss from the fracture. When fractures or vugs are present, typically used additives may not improve fluid-loss control.

Several different fluid-loss additives are available for use with acids. The best additives available today are usually a mixture of a gelling agent that is relatively stable in acid and small inert particles, such as silica flour. The performance of this sytem can be improved when used in an acid-external emulsion fluid. Other often used additives include finely ground acid-swellable polymers and mixtures of finely ground oil-soluble resins. All additives now available must be used in high concentrations (from 100 to 200 lb/1,000 gal) to be effective. A more detailed discussion of these additives is included in Chapter 11.

Effect of Fracture Width

An increase in fracture width normally will increase the distance reactive acid will penetrate along the fracture, the acid penetration distance. This effect is illustrated in Fig. 5.7. For this example, an increase in width from 0.05 to 0.20

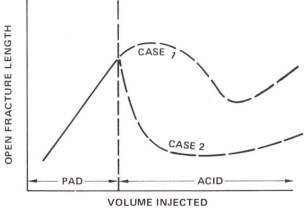

Fig. 5.5—Limits on acid penetration distance.

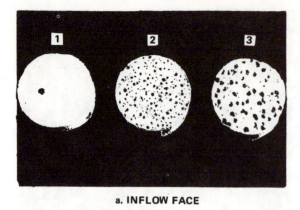

a. INFLOW FACE

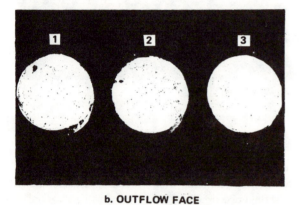

b. OUTFLOW FACE

Fig. 5.6—Cores showing changes in fluid-loss characteristics when fluid-loss additives are used.[5]

in. would increase the acid penetration distance from 80 to 175 ft in a limestone.[4] If the formation is a dolomite, the corresponding increase is from 100 to 260 ft. Conversely, the penetration distance would decrease from 175 to 80 ft if the treatment involved the use of a pad fluid to create a fracture 0.2 in. wide, but the fracture closed to a width of 0.05 in. because a fluid-loss additive was not used in the acid.

Effect of Injection Rate

Altering the injection rate will affect an acid fracturing treatment. The distance reactive acid will penetrate along a fracture normally will increase as the flow velocity along the fracture is increased. An increase in injection rate also will reduce the temperature at which acid enters the fracture, thereby further increasing the acid penetration distance by reducing the reaction rate.

Changes in acid penetration distance with injection rate for example formations, assuming a constant temperature, are illustrated in Fig. 5.8.[4] In this example, the penetration distance approaches 280 ft for a limestone and 400 ft for a dolomite, when the injection rate is about 1 bbl/min per foot of fracture height. Note that the increase in acid penetration distance with an increase in injection rate is lower at high rates. This occurs because the advantage of a higher flow velocity along the fracture is offset by the increased mixing that occurs as a result of the flow.

Effect of Temperature

An accurate estimate of the fluid temperature in a fracture is needed to predict fracture geometry and acid penetration distance. It normally is not accurate to assume the fluid is at formation temperature after it enters the fracture when large-volume, high-rate treatments are considered

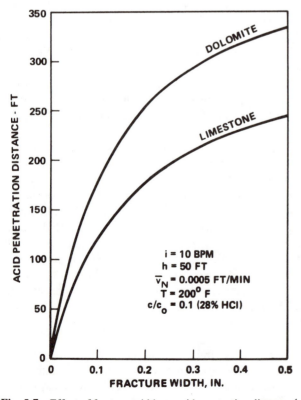

Fig. 5.7—Effect of fracture width on acid penetration distance.[4]

$i = 10$ BPM
$h = 50$ FT
$\bar{v}_N = 0.0005$ FT/MIN
$T = 200°$ F
$c/c_o = 0.1$ (28% HCl)

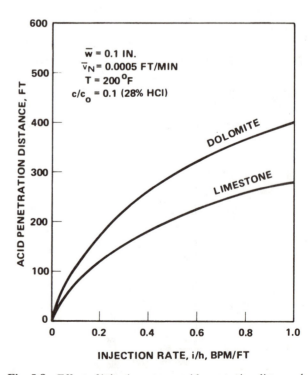

$\bar{w} = 0.1$ IN.
$\bar{v}_N = 0.0005$ FT/MIN
$T = 200°$F
$c/c_o = 0.1$ (28% HCl)

Fig. 5.8—Effect of injection rate on acid penetration distance.[4]

because fluid can reach the perforations at a temperature nearly the same as the surface-injection temperature. Therefore, it is necessary to estimate the change in fluid temperature as it is pumped down the wellbore and along the fracture.

The effect of injection time and rate on temperature at the formation is shown for a 10,000-ft well in Fig. 5.9. The calculations show that a stabilized bottom-hole temperature is approached rapidly.[6] After 1 hour of injection, for example, the bottom-hole temperature is within 50°F of the injection temperature at an injection rate of 4 bbl/min. (At a more typical injection rate of 10 bbl/min, the bottom-hole temperature is within 30°F of the injection temperature after 30 minutes of fluid injection.)

Because of differences in heat capacity and fluid-loss characteristics, the temperature within a fracture will depend on the type of pad fluid used. The effects of fluid type, efficiency (ratio of fracture volume to injected volume), volume injected, and fracture width on the predicted temperature profiles were studied by Sinclair.[7] He concluded the following:

• Fluids with a low viscosity and low efficiency (high rate of fluid loss from the fracture to the formation) can effectively cool the formation near the fracture and thereby reduce fluid temperature in the fracture. Because of the low efficiency, large volumes of fluid will be required to form a long fracture and the fracture width normally will be narrow. Fluids in this category include water alone and low-viscosity gelled waters without fluid-loss additives. These fluids are sometimes used as precooling pads to reduce temperature in the fracture and increase acid penetration distance.

• Viscous fluids normally will have a high efficiency, and therefore less fluid is lost to the formation to cool the rock surrounding the fracture. When these fluids are used, the temperature of the fluid in the fracture will increase and is approximately equal to the formation temperature a short distance from the wellbore.

The effect of average acid temperature in the fracture on acid penetration distance is illustrated for an example treatment in Fig. 5.10. Because the reaction rate at the fracture wall is very fast with limestones (see Chapter 4), reaction is

mass-transfer limited, and the acid penetration distance is essentially independent of temperature. If the formation reacts slowly with the acid, the penetration distance will vary more with temperature. Reasons for these variations are discussed in the next section.

Effect of Formation Type

The reaction rate at the fracture wall is a complex function of rock composition, reservoir temperature, pressure, etc., as discussed in Chapter 4. However, the following general statements can be made about the effect of formation composition on acid penetration distance.

1. Acid reacts with a dolomite more slowly than with a limestone. Therefore, as shown in Fig. 5.10, at low formation temperatures the acid penetration distance will be greater for a dolomite than a limestone. The importance of temperature is clearly illustrated by the decrease in penetration of 28-percent HCl in the dolomite from about 350 ft at 100°F to 175 ft at 220°F. This reduction occurs because the surface reaction rate increases with temperature and at a sufficiently high temperature the reaction is controlled by mass transfer. The reaction rate for the limestone is mass-transfer limited at all temperatures considered. Note that once the reaction is limited by the rate of acid transfer to the fracture walls, there is little difference between the penetration distance in either carbonate.

2. Formations are seldom either a pure dolomite or a pure limestone, as noted in Chapter 3. When a mixture of these carbonates is present, the acid penetration distance will be between predictions for the pure components.

3. If the carbonate formation contains sand or other elements not reactive with hydrochloric acid, tests should be conducted on formation cores to determine the effective surface kinetic parameters.

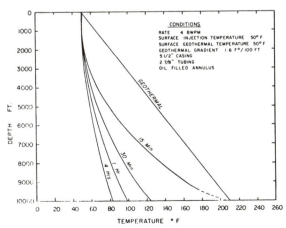

Fig. 5.9—Temperature depth profiles at specified time intervals after injecting cold water at a rate of 4 bbl of water per minute.[6]

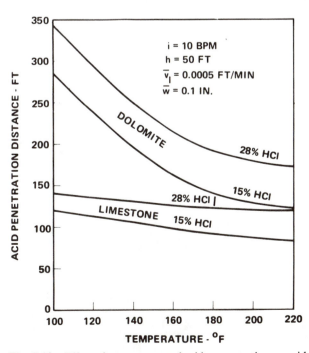

Fig. 5.10—Effect of temperature and acid concentration on acid penetration distance.[4]

TABLE 5.1—VISCOUS ACID COMPOSITIONS

Acid Type	Description
AE1	Acid-external emulsion, consisting of 2 parts kerosene to 1 part 28-percent HCl, containing 2 lb guar per barrel of acid, and 1 percent proprietary emulsifier.
OE1	Oil-external emulsion, consisting of 1 part kerosene to 2 parts 28-percent HCl with 4 percent (by oil volume) duodecylbenzene sulphonic acid emulsifier.
OE2	Oil-external emulsion, consisting of 1 part kerosene to 2 parts 28-percent HCl plus 0.5 percent (by oil volume) proprietary emulsifier.
Gelled HCl	15-percent HCl containing 50 lb of guar per 1,000 gal of acid.

4. Hydrochloric acid does not react to any appreciable extent with sandstone, chert, or other siliceous minerals. Acid fracturing in these formations normally will be unsuccessful, unless reactive carbonates are interbedded with the sand or chert.

Effect of Acid Type and Additives

If the fluid-loss rate of an acid can be controlled, it is sometimes possible to use a retarded acid to maximize the distance acid penetrates along the fracture before being reacted completely. By definition, acids are retarded for acid fracturing purposes only if their reaction rate during flow along the fracture is significantly lower than the reaction rate of the HCl alone. Acids reported to be retarded were tested by Nierode and Kruk[8] under laboratory conditions closely simulating reaction in a fracture. Of the systems tested, they concluded that the best acid for use when a retarded acid is needed is an emulsified system — either acid in oil or oil in acid. Those additives tested that give retardation by coating the fracture surface were found to be useful only at low flow velocities.

Viscous acid systems include emulsified acids and acids gelled with guar or other polymers. Fig. 5.11 illustrates the behavior of some of these acid systems during flow along a fracture. Included in this comparison are gelled 28-percent HCl, two oil-external emulsified acids, and an acid-external emulsified acid. (The compositions of these acids are given in Table 5.1.)

The retardation provided by emulsified acids is primarily a result of the high emulsion viscosity, which tends to reduce the mass-transfer rate to the fracture wall. Shielding the fracture surface with an oil film also may help reduce the reaction rate. Data in Fig. 5.11 show that both the oil-external and acid-external emulsions are retarded and penetrate significantly farther than the gelled hydrochloric acid. Note that after penetrating about 45 ft, Emulsion OE2 broke and the subsequent reaction rate (given by the slope of the concentration-distance plot) was essentially that of the HCl alone.

The oil-external emulsions described in Table 5.1 have a larger dissolving power than the acid-external emulsion. This occurs because the oil-external emulsions tested were about two-thirds 28-percent HCl by volume, whereas the acid-external emulsion contained only one-third 28-percent HCl by volume. The oil-external emulsions therefore have twice the dissolving power. (See Chapter 3 for an extended discussion of the importance and definition of dissolving power.)

If a retarded acid is to be used, it is particularly important to consider the volume of rock dissolved by the acid to assure adequate fracture conductivity is formed. Often, very large volumes of an emulsion may be required. (See Section 5.4 for a discussion of fracture conductivity.)

Gelled acids are commonly prepared by adding polymers such as guar, gum karaya, or a polyacrylamide to HCl. The resulting viscous acid is retarded so long as the fluid is viscous. Unfortunately, retardation in gelled acids is quickly lost because the gelling agent degrades with time and temperature. Fig. 5.12 shows that the viscosity of HCl gelled with 50 lb guar/1,000 gal acid is less than 10 cp after 30 minutes at 100°F. A similar test conducted at 150°F is difficult to interpret quantitatively because the polymer degraded before the samples reached test temperature. When the bottom-hole temperature is less than 150°F, guar or gum karaya added to HCl ranging from 50 to 100 lb/1,000 gal

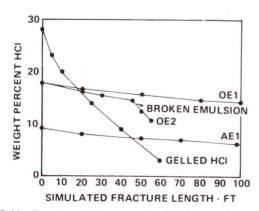

Fig. 5.11—Experimentally measured acid-concentration profiles along a fracture for HCl and emulsified acids.[8]

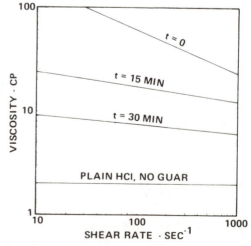

Fig. 5.12—Viscosity of 15-percent HCl containing 50 lb guar/ 1,000 gal acid at 100°F.[8]

should provide some fluid-loss control and retardation; however, the transient nature of the system often makes its use undesirable from a field-operation and cost-effectiveness viewpoint.

Chemically retarded acids, such as those containing the commercially available oil-wetting surfactants tested by Nierode and Kruk,[8] do not retard the reaction rate of HCl under normal acid fracturing field conditions. These additives reportedly function by forming a thin oil film over part of the fracture face, thus shielding some of the fracture area from reaction. When tested under static conditions or at low flow velocities, three such additives gave significant retardation. Fig. 5.13 indicates that the effective diffusion coefficient, which correlates the observed effluent concentrations, is much smaller at low flow rates than for the HCl-limestone reaction without a retarding agent. At typical field flow rates (Reynolds numbers greater than 2,000), however, the reaction rate is equivalent to that for plain HCl. (The definition of the effective diffusion coefficients discussed above is presented in Chapter 4, and experimental values for typically used acids are presented in Chapter 6, Section 6.5.)

Organic acids, such as acetic and formic acids and mixtures of these acids with hydrochloric acid, often have been proposed for use as retarded acid systems.[10-12] The recommended use of organic acids has typically been based on static reaction test data, such as that reported by Dill[11] (Fig. 5.14). Fig. 5.14 was originally interpreted to show that the time required to achieve total acid reaction for the mixed system is longer than for hydrochloric acid alone. Actually, these data illustrate that under reservoir conditions, acetic acid reacts until it is about 50 percent reacted, at which time equilibrium is reached and reaction stops (as discussed in Chapter 3).

The surface reaction rate for the organic acids is lower than for hydrochloric acid, but is normally larger than the rate of acid transfer to the surface at elevated reservoir temperatures. The penetration distance for these acids, therefore, will be similar to conventional hydrochloric acid because the reaction of either acid will be controlled by the rate of acid transfer to the fracture wall by mixing during flow along the fracture.

Also, mixtures of HCl and organic acids are of little benefit because at reservoir conditions CO_2 created by HCl reaction will essentially prevent reaction of the organic acid. These conclusions have been substantiated by Smith *et al.*,[13] who showed that when static reaction-rate data are compared on the basis of reactive acid present in the solution, the organic acids should not have penetration distances appreciably different from hydrochloric acid.

5.4 Fracture Conductivity

To be effective, the acid must react with the walls of the fracture to form a channel that will stay open after the fracturing treatment. Flow channels can be formed as a result of uneven reaction with the rock surface or preferential reaction with minerals heterogeneously placed in the formation, as shown in Fig. 5.15. The conductivity of the fracture is influenced by the volume of rock dissolved (sometimes considered as an acid contact time), the rock strength, and the force attempting to close the fracture (the closure stress).[5]

The effect of volume of rock dissolved, or acid contact time, is shown conceptually in Fig. 5.16. If the contact time is short, the amount of rock dissolved may be insufficient to prevent fracture closure. If excess acid is used, it is possible to reach a situation where the area that must withstand closure forces will fail when stress is applied. The ideal reaction time will dissolve the maximum amount of rock without destroying support for closure stresses. It is, however, important to recognize that each point on the fracture surface is not contacted by acid for the same period of time. This results in a variation of the amount of rock dissolved, and the final fracture conductivity is a function of fracture length. This complex situation is described in Chapter 6, Section 6.6.

The effect of rock embedment strength* and closure stress

---

*Rock embedment strength is defined as the force required to push a ⅛-in.-diameter sphere 1/16 in. into the rock.

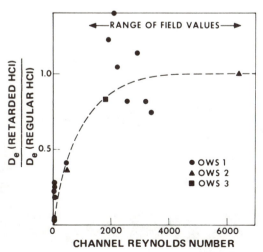

Fig. 5.13—Mixing coefficients for oil-wetting surfactants.[8]

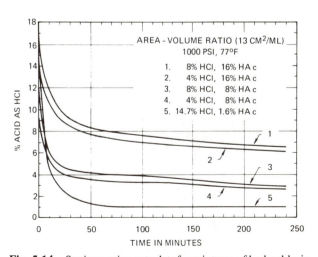

Fig. 5.14—Static reaction-rate data for mixtures of hydrochloric acid and acetic acid.[11]

a. Conductivity created by uneven reaction.

b. Conductivity created by removal of secondary deposit.

Fig. 5.15—Examples of acid reaction patterns that can create fracture conductivity.

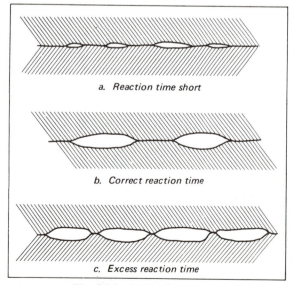

a. Reaction time short

b. Correct reaction time

c. Excess reaction time

Fig. 5.16—Acid etch patterns.

on the fracture conductivity is shown in Fig. 5.17 for conditions where the ideal conductivity, wk_{fi}, would be 10^8 md-in. This figure shows the actual conductivity can be less than 10^4 md-in. if the rock embedment strength is less than 4.5×10^4 psi and the closure stress is 5,000 psi. Here, the ideal conductivity is the conductivity expected if reaction is uniform on the fracture walls. A model for predicting the fracture conductivity from a simple rock embedment-strength measurement has been developed by Nierode and Kruk.[8] This model and a mathematical definition for the ideal conductivity are given in Chapter 6, Section 6.6.

Another mechanism giving rise to irregular surface erosion by the acid is that of "viscous fingering," in which the acid channels through a viscous pad fluid as depicted in Fig. 5.18. In this figure, acid has been injected into a horizontal model of a fracture and is displacing a viscous gelled fluid (appears white in photograph). Channeling will also restrict the portion of the fracture accepting acid, thereby increasing the acid flow velocity.[9] The effect of this increased velocity on penetration distance can be estimated from data such as those plotted in Fig. 5.8. If, for the example considered in Fig. 5.8, the acid channel extends over one-third the fracture height and the average injection rate for the total interval is 0.2 bbl/min/ft, an increase in acid penetration distance from 130 to 210 ft would be expected in the limestone formation.

References

1. Geertsma, J. and de Klerk, F.: "A Rapid Method of Predicting Width and Extent of Hydraulically Induced Fractures," *J. Pet. Tech.* (Dec. 1969) 1571-1581; *Trans.*, AIME, **246**.

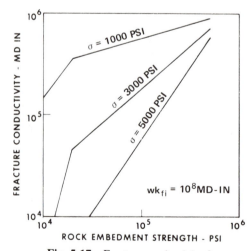

Fig. 5.17—Fracture conductivity.[8]

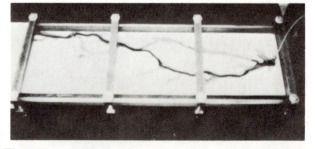

Fig. 5.18—Acid displacement of a viscous non-Newtonian fluid.

2. Williams, B. B. and Nierode, D. E.: "Design of Acid Fracturing Treatments," *J. Pet. Tech.* (July 1972) 849-859; *Trans.,* AIME, **253**.

3. Roberts, L. D. and Guin, J. A.: "The Effect of Surface Kinetics in Fracture Acidizing," *Soc. Pet. Eng. J.* (Aug. 1974) 385-395; *Trans.,* AIME, **257**.

4. Nierode, D. E., Williams, B. B., and Bombardieri, C. C.: "Prediction of Stimulation from Acid Fracturing Treatments," *J. Cdn. Pet. Tech.* (Oct.-Dec. 1972) 31-41.

5. Broaddus, G. C. and Knox, J. A.: "Influence of Acid Type and Quantity in Limestone Etching," paper 851-39-I presented at the API Mid-Continent Meeting, Wichita, Kans., March 31-April 2, 1965.

6. Eickmeier, J. R. and Ramey, H. J., Jr.: "Wellbore Temperature and Heat Losses During Production or Injection Operations," paper 7016 presented at the CIM 21st Annual Technical Meeting, Calgary, May 6-8, 1970.

7. Sinclair, A. R.: "Heat Transfer Effects in Deep Well Fracturing," *J. Pet. Tech.* (Dec. 1971) 1484-1492; *Trans.,* AIME, **251**.

8. Nierode, D. E. and Kruk, K. F.: "An Evaluation of Acid Fluid Loss Additives, Retarded Acids, and Acidized Fracture Conductivity." paper SPE 4549 presented at the SPE-AIME 48th Annual Fall Meeting, Las Vegas, Sept. 30-Oct. 3, 1973.

9. Graham, J. W., Kerver, J. K., and Morgan, F. A.: "Method of Acidizing and Introducing a Corrosion Inhibition Into a Hydrocarbon Producing Formation," U.S. Patent No. 3,167,123 (Jan. 26, 1965).

10. Chamberlain, L. C. and Boyer, R. F.: "Acid Solvents for Oil Wells," *Ind. and Eng. Chem.* (1939) **31**, 400-406.

11. Dill, W. R.: "Reaction Times of Hydrochloric-Acetic Acid Solution on Limestone," paper presented at the ACS 16th Southwest Regional Meeting, Oklahoma City, Dec. 1-3, 1960.

12. Knox, T. A., Pollock, R. W., and Beecroft, W. H.: "The Chemical Retardation of Acid and How it Can Be Utilized," *J. Cdn. Pet. Tech.* (Jan.-March 1965).

13. Smith, C. F., Crowe, C. W., and Wieland, D. R.: "Fracture Acidizing in High Temperature Limestone," paper SPE 3008 presented at the SPE-AIME 45th Annual Fall Meeting, Houston, Oct. 4-7, 1970.

Chapter 6

Acid Fracturing Treatment Models

6.1 Introduction

To design an acid fracturing treatment accurately, it is important to have a model that adequately describes each step in the process. These steps are (1) the fracture geometry, (2) fluid temperature, (3) the acid penetration distance along the fracture for conditions predicted in Steps 1 and 2, and (4) the flow capacity of the fracture created by acid reaction with the walls of the fracture. In this chapter, each topic is discussed, models for each step in the process are compared, and representative models are identified.

6.2 Dynamic Fracture Geometry

General Discussion

Several improved techniques to predict the geometry of vertical fractures have been developed since publication of the monograph, *Hydraulic Fracturing*.[14] These models differ primarily in their assumptions as to how a fracture terminates in the vertical direction. The most common assumption is that the fracture width is uniform over the vertical extent of a formation.[1-4] Another frequently used assumption is that a fracture is long relative to its vertical height.[5-7] The first assumption leads to the prediction that the fracture width is constant over the total vertical height and approximately elliptical along its length. The second assumption gives an approximately elliptical geometry over the vertical height with parallel walls along the fracture length.

Models that assume the fracture width is independent of fracture height predict that width is controlled only by the fracture length, formation elastic properties, and the propagation pressure (average pressure in the fracture generated by fluid flow along the fracture). The accuracy of these models is governed primarily by how the fracture really closes in the overburden and underburden. If the bounding material is deformed easily and does not terminate fracture growth near the interface between the zones, these models are reasonably accurate. When very thin zones are fractured or when the fracture length-to-height ratio is large, these models predict the fracture will be shorter and wider than it actually is in the field.

Models that assume the fracture is long relative to its vertical height predict the width is essentially independent of fracture length and is controlled primarily by the fracture height, formation elastic properties, and the propagation pressure. The accuracy of these models is limited by the assumption that the fracture is contained within a homogeneous section and, therefore, that the elastic properties of the bounding strata are the same as the formation elastic properties. As a result, these models predict the fracture will be longer and narrower than it actually is in the field. At early times during fracture propagation, when the ratio of the fracture length to vertical height is small, these models should predict the fracture is wider than expected in the field.

Neither of the two most commonly used fracture models can always describe accurately what occurs in practice. Limitations exist because each model considers the interval to be homogeneous and isotropic, and neither model includes bounding formations in which the fracture usually terminates. Either model can be used to design fracturing treatments, however, so long as the user recognizes the assumptions and limitations involved in deriving the model.

The model derived by assuming fracture width is constant over the vertical interval to be fractured (hereafter called an unbounded fracture) is used exclusively in this monograph because it is widely used in industry. The use of this model is not intended to imply that acceptable or, in some cases, better results cannot be obtained with other models.

Prediction of Fracture Geometry

Model Development. To predict the physical dimensions of a fracture during a hydraulic fracturing treatment, these dimensions must be related to properties of the formation and the fracturing fluid. Most predictive models therefore are developed by combining analytical solutions to three interdependent problems that describe fracture development when solved simultaneously. This includes equations that describe the following.

1. The Equilibrium Fracture Geometry. These equations relate the fracture length and width to fracture volume, Young's modulus and Poisson's ratio for the formation rock, the statically equivalent pressure in the fracture, and

the formation stress that must be overcome to produce the fracture.

2. Fracture Volume. Equations relating the volume of fluid lost to the formation to fluid and formation properties predict the fracture volume for a known fracture length or surface area.

3. Average Pressure in the Fracture. The force holding the fracture open is generated by the average frictional resistance to flow of the fracturing fluid along the fracture. This pressure is predicted using an equation that relates the pressure gradient attending fluid flow to the fracturing fluid viscosity, the fluid flow velocity, and the fracture width and length.

Note that the fracture geometry is needed to predict the pressure; however, the geometry depends on the average pressure. This interdependence complicates the solution for fracture geometry.

To develop a fracture geometry model, these equations (detailed in Section 6.3) are solved simultaneously using numerical techniques,[2-4] or they are solved analytically after making appropriate simplifying assumptions.[1] A numerical solution is normally preferred because the fracturing fluid characteristics can be altered along the fracture to account for increases in fluid temperature and changes in shear rate.[8] The analytical solution does not allow the fluid viscosity to vary with position, so the user must estimate the average viscosity before calculating the geometry. In many instances, the appropriate average viscosity is difficult to predict.

The equations and development details of a computer model have been discussed by Daneshy,[3] Kiel,[4] Sinclair,[8] and Baron et al.[2] The reader should consult these references for further detail. Similar information for the analytical model is presented by Geertsma and de Klerk.[1] (During the final stages of editing this book, Geertsma and Haafkins presented a paper at the Seventh World Petroleum Congress in Mexico City that compares these predictive techniques — this presentation should be referred to by the reader desiring a detailed comparison of models.)

In the remainder of this section, the design chart developed by Geertsma and de Klerk is discussed and its use is illustrated. This model was included here because it allows *a reasonably accurate prediction of fracture geometry without a computer* and can therefore illustrate factors important in geometry prediction. However, *a numerical solution should be used when available.*

Geertsma and de Klerk Model. Geertsma and de Klerk derived their fracture geometry model by simultaneously solving equations that relate (1) the ratio of the fracture width at the wellbore to the length of one wing of the fracture, and (2) the fracture length to formation and fluid properties.* To simplify the solution of these equations, the results were combined in the design chart given as Fig. 6.1. This chart relates the dimensionless fracture width, K_u, to three dimensionless parameters, $K_{\eta L}$, K_L, and K_s, defined in Eqs. 6.1 through 6.4.

*A similar set of curves also was developed for treatment design when a horizontal fracture is expected.[1]

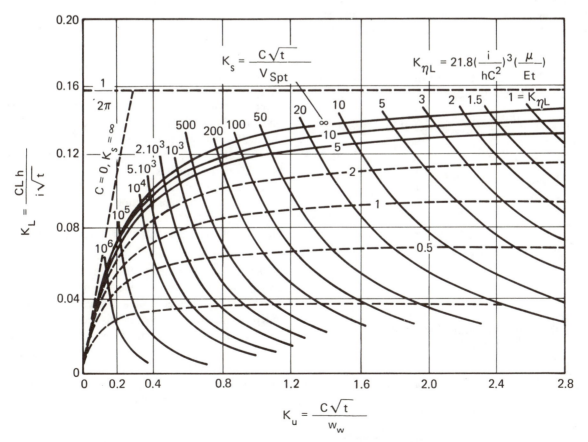

Fig. 6.1—Fracture design chart for linear vertical fracture, Geertsma and de Klerk.[1]

$$K_L = \frac{CLh}{i\sqrt{t}} \quad \dots\dots\dots\dots\dots\dots (6.1)$$

$$K_u = \frac{C\sqrt{t}}{w_w} \quad \dots\dots\dots\dots\dots\dots (6.2)$$

$$K_s = \frac{C\sqrt{t}}{V_{spt}} \quad \dots\dots\dots\dots\dots\dots (6.3)$$

$$K_{\eta_L} = 21.8 \left(\frac{i}{hC^2}\right)^3 \left(\frac{\mu}{Et}\right) \quad \dots\dots\dots\dots (6.4)$$

The following terms are introduced in these dimensionless groups:

C = over-all fluid-loss coefficient (see Eq. 6.27)
E = Young's modulus for the formation rock
h = fracture height
i = fluid injection rate
t = total time for fluid injection
V_{spt} = the spurt volume, defined as the volume of fluid per unit area rapidly lost to the reservoir when new fracture area is created (see Eq. 6.22)
w_w = the fracture width at the wellbore
μ = the viscosity of the fracturing fluid at the temperature existing during flow along the fracture.

An example illustrating the use of Fig. 6.1 is presented next. The selection of input data required to design field treatments is detailed in Chapter 7.

Example Calculation

To illustrate the use of the design curves developed by Geertsma and de Klerk, consider an example treatment on a formation with formation properties and treatment variables as outlined in Table 6.1.

The following groups needed to evaluate K_{η_L} can be derived from the data given in Table 6.1 (note that dimensions must be converted to a consistent set of units):

$$K_s = \frac{C\sqrt{t}}{V_{spt}} = \left[\frac{0.002 \text{ ft}/(\min)^{0.5}}{9.35 \times 10^{-4} \text{ ft}}\right] \sqrt{t}(\min)$$

$$= 2.14 \sqrt{t}. \quad \dots\dots\dots\dots\dots\dots (6.5)$$

$$\frac{i}{h} = \frac{10 \text{ bbl/min}}{50 \text{ ft}} \times \frac{5.615 \text{ cu ft}}{\text{bbl}} = 1.12 \frac{\text{sq ft}}{\min} .$$

$$\dots\dots\dots\dots\dots\dots\dots\dots\dots\dots (6.6)$$

$$\mu = 60 \text{ cp} \times 6.72 \times 10^{-4} \frac{\text{lbm}}{\text{ft sec-cp}} \times 60 \frac{\text{sec}}{\min} ,$$

$$\mu = 2.42 \frac{\text{lbm}}{\text{min-ft}} . \quad \dots\dots\dots\dots\dots (6.7)$$

TABLE 6.1—EXAMPLE FRACTURING TREATMENT — LIMESTONE FORMATION

Formation Properties

Vertical fracture height, h	50 ft
Young's modulus, E	6.45×10^6 psi

Treatment Parameters

Injection rate, i	10 bbl/min
Fluid viscosity, μ	60 cp
Fluid spurt, V_{spt}	0.00935 cu ft/sq ft
Over-all fluid-loss coefficient, C	0.002 ft/min $^{1/2}$

$$E = 6.45 \times 10^6 \text{ psi} \times 4.63 \times 10^3 \frac{\text{lbm}}{\text{ft sec}^2\text{-psi}}$$

$$\times (60)^2 \frac{\text{sec}^2}{\min^2} ,$$

$$E = 1.08 \times 10^{14} \frac{\text{lbm}}{\text{ft-min}^2} . \quad \dots\dots\dots\dots (6.8)$$

$$K_{\eta_L} = 21.8 \left[1.12 \frac{\text{sq ft}}{\min} \times \left(\frac{1}{0.002 \text{ ft/min}^{1/2}}\right)^2\right]^3 \times$$

$$\left[\frac{2.42 \text{ lbm} - \text{ft-min}^2}{\min\text{-ft} \times 1.08 \times 10^{14} \text{ lbm}} \frac{1}{t}\right],$$

$$K_{\eta_L} = \frac{21.8 \,(2.22 \times 10^{16})(2.25 \times 10^{-14})}{t}$$

$$= \frac{1.08 \times 10^4}{t} , \quad \dots\dots\dots\dots\dots (6.9)$$

where t is now in minutes.

By rearranging Eqs. 6.1 and 6.2 the fracture length and wellbore width can be related to K_L and K_u by Eqs. 6.10 and 6.11.

$$L = K_L \left(\frac{i\sqrt{t}}{hC}\right) ,$$

$$= \left(1.12 \frac{\text{sq ft}}{\min} \times \frac{1}{0.002 \text{ ft/min}^{0.5}}\right) K_L \sqrt{t},$$

$$L = 560 \, K_L \sqrt{t}, \text{ ft.} \quad \dots\dots\dots\dots\dots (6.10)$$

$$w_w = \frac{C\sqrt{t}}{K_u} ,$$

$$w_w = 12 \text{ in./ft} (0.002 \text{ ft/min}^{0.5}) \frac{\sqrt{t}}{K_u} ,$$

$$w_w = \frac{0.024\sqrt{t}}{K_u} , \text{ in.} \quad \dots\dots\dots\dots\dots (6.11)$$

The fracture geometry as a function of time is determined by (1) substituting the time of interest into Eqs. 6.5 and 6.9, (2) entering Fig. 6.1 for the calculated values of K_s and K_{η_L}, (3) reading values for K_u and K_L, and (4) calculating the fracture width and length with Eqs. 6.10 and 6.11. This calculation is summarized in Table 6.2 for times of 15, 30, 45, and 60 minutes. The predicted fracture width and length can be plotted vs time on log-log paper to obtain a continuous geometry prediction.

6.3 Equations Used to Define Fracture Geometry

It is necessary to solve simultaneously equations that predict fracture geometry, fracture volume, and average pressure in the fracture, to develop a model like the one just discussed. The detailed equations used to model each factor independently are reviewed in this section. This discussion will acquaint the practicing engineer with the important variables that appear in the relationships and are critical for an accurate prediction of fracture geometry. The limitations of these relationships and proper definitions for use in the equations will be discussed also.

Equilibrium Fracture Geometry

Expressions for fracture width and length for an un-

TABLE 6.2—PREDICTION OF FRACTURE GEOMETRY WITH PROCEDURE
OF GEERTSMA AND DE KLERK

Time (minutes)	K_s	K_{η_L}	K_L	K_u	Length (ft)	Wellbore Width (in.)	Average* Width (in.)
15	8.29	720	0.108	0.58	235	0.16	0.13
30	11.72	360	0.116	0.70	357	0.19	0.15
45	14.35	240	0.120	0.75	453	0.21	0.17
60	16.58	180	0.123	0.81	536	0.23	0.18

*This is determined using Eq. 6.15 given in the following section.

bounded vertical fracture were derived by Khristianovich and Zheltov[10,11] and expanded by Barenblatt[12] as follows.

$$w = \frac{4(1-\nu^2)\,\bar{p}_f L}{\pi E} \left\{ (\cos\theta)\ln\left[\frac{\sin|\theta-\theta_p|}{\sin(\theta+\theta_p)}\right] + (\cos\theta_p)\ln\left(\frac{\tan\frac{\theta+\theta_p}{2}}{\tan\frac{\theta-\theta_p}{2}}\right) \right\}, \quad \dots\dots (6.12)$$

where $x = \cos\theta$, $\dots\dots\dots\dots\dots\dots\dots$ (6.13)

$$x_p = \cos\frac{\pi}{2}\left(1 - \frac{\sigma}{\bar{p}_f}\right) \quad \dots\dots\dots\dots\dots (6.14)$$

To obtain this solution, they assumed that the fracture is generated by a constant pressure, $\bar{p}_f$, acting over a fraction of the total fracture length, x_p, called the pressurized fracture length, as illustrated in Fig. 5.1. The remainder of the fracture is unpressurized. The closure or for field stress is σ.

The average width of a fracture can be obtained by integrating Eq. 6.12 over the fracture length. The average width, $\bar{w}$, can be related to the width at the wellbore, w_w, by Eq. 6.15 as

$$\bar{w} = \frac{\pi\,w_w}{4}. \quad \dots\dots\dots\dots\dots\dots\dots\dots (6.15)$$

The fracture length can be related to fracture volume and rock properties using Eq. 6.16.*

$$L = \left[\frac{V_f E}{2(1-\nu^2)h\,\bar{p}_f \sin(\pi\sigma/\bar{p}_f)}\right]^{0.5} \quad \dots\dots (6.16)$$

The geometry predicted by Eq. 6.12 is accurate only if the average pressure is chosen so that the force and moment generated by the constant pressure is equal to that generated by the actual pressure distribution. Daneshy[3] has shown that the geometry prediction can be improved if it is obtained using a pressure distribution other than the square distribution. He proposed the general form $p(x) = \bar{p}_f\sqrt{1-(x/x_p)^n}$. This modification is normally of relatively small importance; therefore, Eq. 6.12 is usually satisfactory.

The primary limitations of this fracture geometry model are

1. It assumes the complete vertical section to be fractured opens instantaneously. Therefore, calculations for massive formations will predict a fracture width smaller than the width actually created because it is unlikely that the total section will fracture initially in practice.

2. It does not include restrictions in width when hard-to-

*Private communication from L. E. Goodman (U. of Minnesota).

fracture formations are above or below the section to be fractured. In these cases, the fracture width can be controlled by the characteristics of the bounding strata, and the fracture width predicted using formation properties often will be too large.

3. The width-to-length ratio is dependent on the accuracy with which Young's modulus for the formation rock and the pressure generated by fluid flow along the fracture are known.

Fluid Loss

Control By Formation and Fluid Properties. Howard and Fast[13] postulated in 1957 that three resistances affect the rate of fluid loss from the fracture. These resistances include the filter cake formed on the fracture face when fluid-loss prevention additives are used, invasion of the formation matrix by a viscous fracturing fluid, and compression of reservoir fluids by invasion of fracturing fluids. The expressions, given as Eqs. 6.17 through 6.19, were derived to predict the leakoff velocity, assuming it is totally controlled by a single mechanism.

Compression of reservoir fluid is expressed as

$$v_N = \sqrt{\frac{k\phi\,\kappa_{fl}(\Delta p_c)^2}{\pi\,\mu_f t}} = \frac{C_c}{\sqrt{t}}, \quad \dots\dots\dots\dots (6.17)$$

where κ_{fl} is the isothermal compressibility of the formation fluids, μ_f is the reservoir fluid viscosity, and Δp_c is the pressure drop between the invading fracture fluid and the average reservoir pressure. If the velocity is to be expressed in feet per minute, the factor of π is dropped and the expression is multiplied by 0.0469 for k in darcies, t in minutes, Δp in psi, μ in centipoise, and κ_{fl} in psi^{-1}.

Fracture-fluid invasion of the formation is expressed as

$$v_N = \sqrt{\frac{k\phi\,(\Delta p_v)}{2\,\mu t}} = \frac{C_v}{\sqrt{t}}, \quad \dots\dots\dots\dots\dots (6.18)$$

where Δp_v is the pressure difference between the fracture wall and the interface between the penetrating fracture fluid and formation fluids. If the velocity is to be expressed in feet per minute, the expression is multiplied by 0.0374, the factor of 2 is eliminated, and other parameters are expressed in the units defined above.

Filter cake is expressed as

$$v_N = \frac{C_w}{\sqrt{t}}. \quad \dots\dots\dots\dots\dots\dots\dots\dots\dots (6.19)$$

Each coefficient defined above (C_c, C_v, and C_w) is the fluid-loss coefficient for the particular flow resistance of interest.

These equations describing the flow resistance in the reservoir can be combined by noting that the over-all pressure drop between the fracture and the formation is the sum of the pressure drop in each zone. Introducing the definition for the fluid-loss coefficients C_v and C_c into a pressure balance, redefining C_v and C_c in terms of the over-all pressure drop instead of the pressure drop across the individual zones, defining an over-all fluid-loss coefficient, C_{vc}, so that $v_N = C_{vc}/\sqrt{t}$, and then solving for the over-all fluid-loss coefficient, C_{vc}, we find it is related to the individual zone coefficients by

$$C_{vc} = \frac{2\,C_c C_v}{C_v + \sqrt{C_v^2 + 4 C_c^2 C_v^2}}. \qquad (6.20)$$

Eq. 6.20 is the correct equation for the combined fluid-loss coefficient. The commonly used *harmonic mean* (Eq. 6.21), *therefore, is not a proper equation* for combining these effects.

$$\frac{1}{C_{vc}} = \frac{1}{C_v} + \frac{1}{C_c}. \qquad (6.21)$$

Control by Fluid-Loss Additives. The fluid-loss control characteristics of solid fluid-loss additives can be determined from static or dynamic fluid-loss test procedures (reviewed by Howard and Fast[14]) when fluid loss occurs into the formation matrix.* Normally, the fluid-loss tests are characterized by an initially rapid rate of fluid loss before the cake forms, followed by a low rate of fluid loss after cake formation. The volume of fluid loss per unit area measured in a static test can be characterized by the following equation:

$$V = V_{spt} + 2 C_w \sqrt{t}, \qquad (6.22)$$

and the results from a dynamic test, one where fluid flows past the core surface during the test, are best represented by

$$V = V_{spt} + v_N t. \qquad (6.23)$$

Williams[15] has proposed that the fluid spurt is often not instantaneous so fluid losses before cake formation should be characterized using Eq. 6.24 and thereafter by either Eq. 6.22 or Eq. 6.23.

Thus, for $t < t_{spt}$, where, $t_{spt} = (V_{spt}/2C_{vc})^2$,

$$V = 2 C_{vc} \sqrt{t}. \qquad (6.24)$$

For $t > t_{spt}$, Eqs. 6.22 and 6.23 should be modified as follows

For a static test,

$$V = V_{spt} + 2 C_w \sqrt{t - t_{spt}}, \qquad (6.25)$$

or for a dynamic test,

$$V = V_{spt} + v_N (t - t_{spt}). \qquad (6.26)$$

In these equations, C, the over-all fluid-loss coefficient, is given by the following expression, which can be derived in a manner analogous to the derivation of Eq. 6.20.

$$C = \frac{2\,C_c C_v C_w}{C_v C_w + \sqrt{C_w^2 C_v^2 + 4 C_c^2 (C_v^2 + C_w^2)}}. \qquad (6.27)$$

Neither the dynamic nor static fluid-loss test is completely representative of conditions under which fluid losses occur in the fracture. Flow past the rock surface in the dynamic test restricts filter cake buildup; therefore, this test should be representative of fluid losses over the majority of the fracture length. Near the fracture tip, however, the flow velocity approaches zero and fluid losses occur under conditions similar to those present in a static test. As noted earlier, the fluid-loss rate into naturally occurring fractures or vugs will be much higher than measured in either test.

Integral Fluid-Loss Equations

Geertsma and de Klerk derived an over-all volume balance equation, assuming the fluid-loss model given as Eq. 6.22 applies. Williams derived similar volume balance equations for the assumption that fracture length is proportional to the square root of time.

Propagation Pressure

Fracture geometry is directly related to the magnitude of the average pressure acting along the pressurized portion of the fracture to oppose the minimum far field stress. This pressure is generated by fluid flow along the fracture. For the assumption of no fluid loss, the total pressure drop along a growing fracture for a Newtonian fluid is approximately given by[16]

$$\Delta p_f = \frac{12\mu i L}{h} \int_0^{x_p L} \frac{dx}{w^3}. \qquad (6.28)$$

For the more realistic assumptions of fluid loss from the fracture and a variation of fluid properties along the fracture (because of changes in fluid temperature and shear rate), the pressure drop is given by

$$\Delta p_f = 24 L \int_0^{x_p L} \frac{\mu \bar{v}_A dx}{w^2}. \qquad (6.29)$$

The use of Eq. 6.29 is complicated because the fluid viscosity, flow velocity along the fracture, $\bar{v}_A$, and fracture width all vary with position along the fracture and are interdependent. That is, the fluid viscosity at a position x is a function of the fluid temperature, the fracture width, and the flow velocity at the position in the fracture. Therefore, it is impossible to specify analytic functions for $\bar{v}_A$ and μ to allow direct integration of these equations. An accurate pressure drop prediction that includes both temperature and shear rate effects can be attained, therefore, only by numerical integration.

6.4 Fluid Temperature in the Fracture

The importance of an accurate estimate of the fluid temperature as the fluid enters and moves along the fracture was discussed in Chapter 5. Models available to make these predictions are discussed in the remainder of this section.

---

*Procedures for predicting the fluid-loss rate into natural fractures that may be intersected by the induced fracture are not available. However, it is safe to conclude that fluid loss rates are high and that normally used fluid-loss additives may not be effective.

Bottom-Hole Injection Temperature

Eickmeier and Ramey Model. A useful procedure for predicting bottom-hole fluid temperatures during fluid injection was developed by Eickmeier and Ramey.[17] These authors derived a finite difference model for heat transfer between the injected fluid, and the wellbore and surrounding formation during injection or production operations. The model has particular application to the early transient period when both the rate of heat loss and wellbore temperatures are changing rapidly. It is constructed so that a realistic wellbore model (including tubing, annulus fluid, casing, cement, and formation) can be considered in calculations.

The basic assumptions incorporated into the numerical model are (1) there is no vertical heat transfer by conduction in either the wellbore or the formation, (2) thermal properties are considered to be independent of temperature, (3) the heat-transfer coefficient remains constant, (4) heat transfer to or from the fluid initially in the tubing is not considered, (5) surface injection temperature and injection rate are constant, (6) geothermal temperature may be specified as a linear function of depth, (7) the tubing, casing, and hole size are constant over the entire depth of the well, and (8) friction and kinetic energy are not considered.

A typical temperature profile predicted with this model is given in Fig. 6.2. This profile shows that (1) there is no radial temperature gradient in the tubing (the model assumes complete mixing — that is, turbulent flow), (2) a sharp gradient exists across the oil-filled annulus and cement zones because of the low thermal conductivity of the oil and cement, and (3) there is a gradual increase in temperature within the formation. Note that a significant temperature increase has penetrated a distance of only 15 in. after 4 hours of injection.

Other Models for Predicting Injection Temperature. Other analytical and numerical models for predicting bottom-hole injection temperatures were published before the Eickmeier and Ramey model. In general, the analytical solutions are limited in that they consider only steady state (long injection times) and do not consider the details of the near-wellbore region.

The most important *analytical models* were developed by Moss and White,[18] Ramey,[19] and Squier *et al*.[20] The original paper by Moss and White dealt with the calculation of temperature profiles in wells during fluid injection or production. Their analysis was based on the constant-rate, line-source solution to the diffusivity equation. They assumed that the injection fluid temperature was the same as the casing temperature at any particular depth. To use this approach, it is necessary to divide the well depths into lengths of 100 to 1,000 ft.

Ramey used a similar approach; however, he included the geothermal gradient so that it was not necessary to divide the well depth into incremental lengths. He also simplified the calculations by introducing dimensionless quantities and an over-all heat-transfer coefficient to account for the effects of steady-state wellbore conditions. His results are considered accurate for time periods greater than 1 week for most typical problems.

Squier *et al*. presented an exact, short time solution for temperature behavior of hot-water injection wells and also an asymptotic solution valid for large times. One assumption incorporated into the solution is that the fluid temperature at any depth, and the wellbore formation face temperature at that depth, are the same — that is, the presence of casing or cement is neglected.

Recently, several *numerical models* have been developed to predict mud and cement circulation temperatures that, with slight modification, can be used to predict injection temperature. The most important contributions have been by Leutwyler,[21] Claycomb and Schweppe,[22] Raymond,[23] Keller *et al*.,[24] and Sump and Williams.[25]

Temperature Profile Along the Fracture

The Wheeler Model. The model preferred by the authors for predicting the temperature profile along a vertical fracture was derived by Wheeler.[26] This solution relates the dimensionless temperature to the fracture width, injection rate, time of injection, rate of fluid loss, and the fluid and formation thermal properties. Assumptions used to derive the model are (1) the physical properties of the formation and fluids are independent of temperature, (2) the rate of fluid loss to the formation is constant at all points along the fracture, (3) the fracture width is constant, and (4) heat transfer to the fracture occurs only in the direction perpendicular to the fracture face.

For these assumptions, the dimensionless temperature at any point along the fracture was defined. Because of the complication of the resulting equations, the reader should consult Ref. 26 for details.

Sinclair used Wheeler's model to calculate the effect of heat transfer on fracture-treatment design.[27] His calculations show that, if the fluid injected is cool, its temperature when entering the fracture may be significantly below reservoir temperature. As a result, the fracturing fluid will be more viscous than a fluid instantaneously heated to formation temperature on entering the fracture. As a result, the predicted fracture width will usually be greater than if one assumes the fracturing fluid is at reservoir temperature.

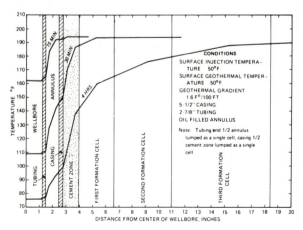

Fig. 6.2—Cross-section temperature profile at 9,000 ft after cold-water injection.[17]

Other Models. The first attempts to include heat-transfer effects in fracturing treatment design[28, 29] considered only heat conduction from the fracture. Because convective heat transfer caused by fluid losses to the formation were not considered, these calculations are of limited value.

In 1969, Whitsitt and Dysart[30] presented a model that accounted for both conductive and convective heat transfer to the formation. To derive this model, they assumed that the rate of fluid loss varied linearly from zero at the wellbore to a maximum at the fracture tip. It is believed that this model is less accurate than Wheeler's model because the assumed distribution of fluid losses does not represent as realistically the processes that occur in fracturing. A comparison of temperature profiles predicted by the two models is given in Fig. 6.3. This figure shows that because higher rates of fluid loss are assumed near the fracture tip, the Whitsitt and Dysart model predicts a lower temperature than the Wheeler model and, therefore, may cause the fracture width to be overestimated.

6.5   Acid Penetration Along the Fracture

A number of experimental and theoretical models have been proposed to predict the distance reactive acid can penetrate along a hydraulically induced fracture. The models believed to be most representative of the acid-fracturing process were developed by Roberts and Guin,[31, 32] Nierode and Williams,[33] Nierode *et al.*,[34] Whitsitt *et al.*,[35] and van Domselaar *et al.*[36] All of these models consider the effects of surface kinetics, flow along the fracture, and fluid loss from the fracture on the acid-reaction process. Only the model proposed by Roberts and Guin will allow the direct inclusion of a surface kinetic expression determined for the formation of interest. Other models include surface kinetic effects by altering the rate of acid transfer to the surface to match experimental results obtained for acid reaction in a simulated fracture created with formation core material.

The approach proposed by Roberts and Guin is sound for highly turbulent flow where it is expected that the mass

transfer rate to the fracture surface can be studied separately from the reaction kinetics. If, as is expected in many field circumstances, flow is not highly turbulent, natural convection (occurring because the density of the products exceeds that of the acid) can control acid transfer to the fracture wall. In this condition, it appears unlikely that the mass transfer mechanism can be studied apart from the kinetics because the convection rate is a function of the rate of reaction. Studies in which the effective mixing coefficient is measured together with the chemical reaction, using a properly oriented fracture system, are probably more reliable, provided the anticipated field leak-off rate is properly modeled in the laboratory. Calculations presented in Chapter 7 use effective mixing coefficients derived empirically by Roberts and Guin and Nierode and Williams from experiments closely simulating field conditions (vertical fracture, rapid reaction, intermediate Reynolds numbers) for the reason just discussed.

Models for Acid Reaction Along the Fracture

The Roberts-Guin Model. To develop their design model, Roberts and Guin solved simultaneously the following three equations:

$$\frac{\bar{w}}{2L} \frac{d\bar{v}_A}{dx} + \bar{v}_N = 0, \qquad (6.30)$$

$$\frac{\bar{w}}{2L} \bar{v}_A \frac{d\bar{c}}{dx} = (\bar{v}_N - K_g)(\bar{c} - c_w), \qquad (6.31)$$

$$K_g (\bar{c} - c_w) = \xi_f' c_w{}^m, \qquad (6.32)$$

where $\bar{v}_A$ is the average flow velocity along the fracture at any position, $\bar{v}_N$ is the average fluid-loss velocity from the fracture, and $\bar{c}$ is the average acid concentration in the fracture at some specified position.

Eq. 6.30 is a modified form of the continuity equation; the fluid loss through the fracture wall, $\bar{v}_N$, results in a decrease in the average velocity along the fracture, $\bar{v}_A$. Eq. 6.31 is an averaged form of the convective diffusion equation relating the change in acid concentration to the flow velocity, fracture width, and the mass-transfer coefficient, K_g. Eq. 6.32 is an expression equating the rate of acid disappearance at the wall by the surface reaction to the rate of acid flux to the wall. ξ_f' and m are reaction rate parameters defined in Chapter 4.

To predict the acid concentration profile along the fracture, these equations are solved simultaneously for the three unknowns, $\bar{v}_A$, $\bar{c}$, and c_w, as a function of distance down the fracture. This solution requires that the mass-transfer coefficient, K_g, and the reaction rate constants, ξ_f' and m, be known.

For the case of first-order kinetics ($m = 1$), a constant fluid loss and a constant mass-transfer coefficient rate, Eqs. 6.30 through 6.32 can be integrated analytically to give the following equations relating the fraction of the injected acid concentration remaining, $\bar{c}/c_o$, to the dimensionless position along the fracture, x.

$$\frac{\bar{c}}{c_o} = \left(1 - \frac{2Lx}{\bar{w}} \frac{N_{Pe}*}{N_{Pe}}\right)^n, \qquad (6.33)$$

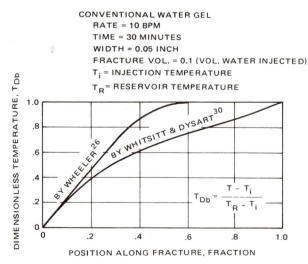

CONVENTIONAL WATER GEL
RATE = 10 BPM
TIME = 30 MINUTES
WIDTH = 0.05 INCH
FRACTURE VOL. = 0.1 (VOL. WATER INJECTED)
T_i = INJECTION TEMPERATURE
T_R = RESERVOIR TEMPERATURE

$$T_{Db} = \frac{T - T_i}{T_R - T_i}$$

BY WHEELER[26]

BY WHITSITT & DYSART[30]

DIMENSIONLESS TEMPERATURE, T_{Db}

POSITION ALONG FRACTURE, FRACTION

Fig. 6.3—Temperature profiles, comparison of Wheeler's constant-leakoff calculation with Whitsitt and Dysart's linearly increasing calculation.[27]

where

$$n = \frac{P_f(N_{Sh} - 4N_{Pe}^*)}{N_{Pe}^*(4P_f + N_{Sh})} \quad \ldots \ldots \ldots \ldots \ldots \quad (6.34)$$

In Eqs. 6.33 and 6.34 and subsequent discussions, the following dimensionless groups are introduced.

D_e = effective diffusion coefficient,
N_{Pe} = Peclet number for flow along the fracture =
  $\bar{w}\bar{v}_{A\circ}/2 D_e$
  where $v_{A\circ}$ is the average velocity at the fracture inlet,
N_{Pe}^* = Peclet number for fluid loss = $\bar{w}v_N/2 D_e$,
N_{Sh} = Sherwood number = $2\bar{w} K_g/D_e$,
N_{Sc} = Schmidt number = $\mu/D_e\rho$,
P_f = dimensionless surface reaction rate = $\bar{w}\xi_f'/2 D_e$
  (for a first-order reaction).

For situations where the reaction is not first order, $m \neq 1$, Eqs. 6.30 through 6.32 must be solved numerically.

Eq. 6.33 relates the acid concentration for first-order reactions ($m = 1$) and for constant leak-off rates to distance measured down the fracture. The contribution of the rate of chemical reaction is embodied in the dimensionless reaction rate, P_f. Experimental methods for its determination are described in Chapter 4. The rate of mass-transfer coefficient is included in the Sherwood number. For Reynolds numbers greater than 5,000, natural convection is not important and the Sherwood number can be obtained from Fig. 6.4 (note that results correlate with the well known Gill[38] and Seider-Tate[37] heat-transfer models). Eq. 6.33 can then be used to determine the acid penetration distance. It should be noted that Fig. 6.4 is strictly applicable if there is no fluid loss from the fracture. However, Roberts and Guin suggest that for most practical situations, this correlation is adequate because the fluid-loss velocity is small relative to the velocity along the fracture.

At small Reynolds numbers, the situation depicted in Fig.

6.5 arises and the Sherwood number is no longer a unique function of the Schmidt number and the Reynolds number. The effective diffusion coefficient should be obtained from Fig. 4.2 when using either Fig. 6.4 or 6.5. For this reason, the contribution of convection to mass transfer is substantial and by orienting the fracture at various angles with respect to the vertical, different rates of mass transfer are observed. These rates also depend to some extent on the reaction rate. In the future, it may be possible to predict the influence of natural convection on the Sherwood number, but this capability does not now exist.

Nierode-Williams Model. In their study of the acid fracturing treatment, Nierode and Williams[33] first determined a reaction rate expression for the reaction of hydrochloric acid at a limestone surface. The resulting reaction rate model shows that for typical field conditions in limestone formations, acid reaction is limited by the rate of and acid transfer to the rock surface, not by the surface reaction rate. The concentration of acid near the surface is, therefore, often quite small and can be assumed equal to zero. This conclusion was not found for all carbonates. For example, the reaction with a dolomite is usually limited by both reaction rate and diffusion.

To predict the reaction of acid during flow along a fracture, Nierode and Williams used a solution given by Terrill[39] for the analogous heat-transfer problem. Van Domselaar *et al.*[36] derived a comparable solution to this problem and obtained identical results when used with the same input data. The Nierode and Williams solution incorporates the following assumptions: (1) laminar, incompressible flow along the fracture, (2) acid viscosity as it reacts while moving along the fracture is constant, (3) reaction rate at the rock surface is infinite; that is, acid concentration at the surface is zero, and (4) constant rate of fluid loss from the fracture to the formation.

To simplify the solution, it is presented in graphical form in Fig. 6.6. This figure allows the mean concentration, $\bar{c}$, to

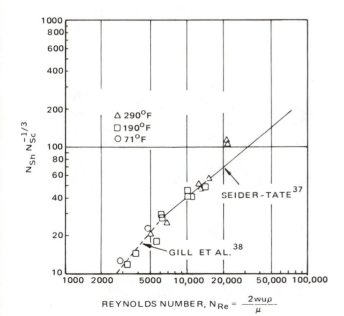

Fig. 6.4—Sherwood number for acid reaction in turbulent flow.[32]

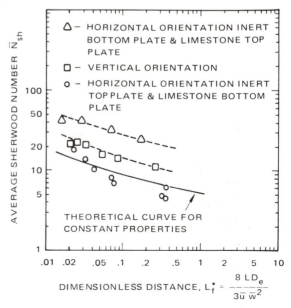

Fig. 6.5—Influence of fracture orientation on mixing.[32]

be specified by fixing values for three dimensionless groups, N_{Re*}, N_{Re}, and N_{Pe*}. These groups are defined as follows:

N_{Re*} = Reynolds number for fluid loss, $2\overline{w}v_N\rho/\mu$,

N_{Re} = Reynolds number for flow along the fracture,
$2\overline{w}\overline{v}_A°\rho/\mu$,

N_{Pe*} = Peclet number for fluid loss, $\overline{w}v_N/2D_e$.

This figure allows the dimensionless position at which the mean acid concentration reaches a desired level, $\overline{c}/c_o$, to be related to the Peclet number. With the exception of the effective mixing coefficient, D_e, the parameters that must be known to predict the acid penetration distance are fixed by formation characteristics or by fracture dimensions. If the fluid-loss velocity is zero, Fig. 6.6 cannot be used. In this case use Figs. 4.5 through 4.8, as appropriate.

The effective mixing coefficient, D_e, is, in effect, an adjustable parameter that must be chosen to allow the mathematical model for acid reaction rate to agree with laboratory experiments designed to simulate accurately acid reaction rate in the field. If acid flow was laminar and acid transfer was by ion diffusion alone, the effective mixing coefficient would be equal to the ionic diffusion coefficient for the acid of interest. The mixing coefficient will normally be larger than the ionic diffusion coefficient, since it will include acid transfer caused by secondary flow. As illustrated in Fig. 5.3, secondary flow can be caused by changes in acid density with reaction, wall roughness, turbulence, or a combination of these effects.

To obtain accurate values for the effective mixing coefficient, acid reaction rate was measured in experiments that closely represented field conditions. The experimental equipment used by Nierode and Williams[34] (Fig. 6.7) was constructed so that reaction rate was measured during acid flow between parallel walls of reactive rock. To assure that the fracture walls were of a surface roughness typical of surfaces formed when a formation is fractured, cores used in the equipment were prepared from a cylindrical core that was fractured in tension. This technique allowed the walls of the test equipment to have very rough surfaces with a uniform fracture width. During the experiments, the fracture was oriented to represent a vertical fracture to allow gravity forces to influence acid mixing properly. The system temperature, pressure, and rate of fluid loss through the fracture walls were set at typically observed levels. Effective mixing coefficients were determined from experiments by fitting the computer solution to equations describing reaction in a parallel plate system to the experimental conditions and the acid concentration ratio, $\overline{c}/c_o$, measured in the experiment.

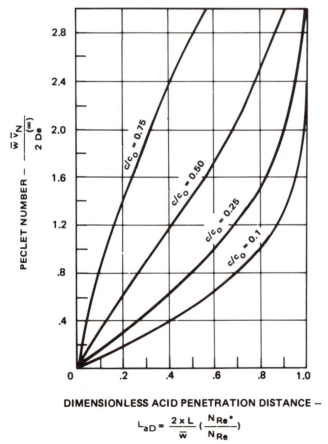

Fig. 6.6—Acid penetration distance along a fracture.[33]

b. Cross Section

Fig. 6.7—Experimental fracture model, Nierode and Williams.[34]

Similar tests were later run by Roberts and Guin.[31]

The combined mixing coefficient data obtained by Williams and Nierode and Roberts and Guin are presented in Fig. 6.8. The $D_e^{(\infty)}$ defined in Fig. 6.8 represents the effective diffusion coefficient used to calculate the acid penetration distance. The superscript (∞) indicates that this coefficient is good only when the reaction rate is fast, relative to the rate of acid transfer to the surface (limestone formations). For Reynolds numbers greater than 5,000, the Roberts and Guin approach described previously (Fig. 6.4) can be used.

The reaction rate of hydrochloric acid at the fracture wall will not necessarily be infinite if the formation to be treated is not a limestone. When surface kinetics are not infinite, the coefficient required to fit experimental data with the theory derived by assuming an infinite reaction rate will reflect the surface kinetic limit. Mixing coefficient data for the reaction of hydrochloric acid and Kasota dolomite (Fig. 6.9) gives the ratio between the effective diffusion coefficient for a dolomite, D_e, where the surface reaction rate affects reac-tion kinetics, and coefficient $D_e^{(\infty)}$ defined in Fig. 6.8. For convenience, smooth-walled cores were used in these experiments, and the effect of wall roughness was added by increasing the mixing coefficient by a factor determined from the experiments with Indiana limestone, which forced mixing coefficients for smooth cores to agree with those measured in the rough-walled fracture. The corrected effective mixing coefficient for hydrochloric acid reaction with Kasota dolomite is a function of temperature and can be related to $D_e^{(\infty)}$ by Eq. 6.35 for temperatures greater than 535°R as

$$\frac{D_e}{D_e^{(\infty)}} = 1 - \exp\left[2{,}445\left(\frac{1}{T} - \frac{1}{508}\right)\right], \quad \ldots \ldots (6.35)$$

where T is in °R.

Mixing coefficients required to model acid reaction with other formation types should follow the general behavior illustrated above. To determine acid reaction characteristics with other formations, it is necessary to run dynamic reaction rate tests with formation cores to obtain effective mixing coefficient data. In general, however, reasonable results can be obtained by choosing mixing coefficients for the rock type most characteristic of the formation — that is, limestone or dolomite. An alternative procedure for a treatment to be conducted with a high Reynolds number would be to measure the rate of surface reaction (see Chapter 4) and then use the Roberts and Guin approach to determine a surface kinetic model.

Because the Nierode and Williams model has been fitted to experiments that closely model fracturing, it should give a reasonable representation of the acid-fracturing process. There are, however, two assumptions involved in derivation of the model that sometimes may limit its use.

1. Acid viscosity is not a function of concentration. This obviously is not correct for very concentrated acid solutions. The use of the average acid viscosity at the temperature and pressure expected in the fracture, however, should allow reasonable predictions to be made.

2. Acid concentration at the fracture wall is zero. This assumption is valid for most limestone formations, but it is probably not correct for most dolomites or mixed composition formations. *The model can be used to simulate a slowly reacting formation only if effective mixing coefficient data have been obtained for that formation or one with a similar surface reaction rate.*

When the model is used under conditions where the surface reaction rate is controlling, it must be considered as a phenomenological model with little theoretical basis. The problem of acid reaction in fracturing treatments may be solved for the assumption of finite reaction rate at the fracture wall, using procedures proposed by Roberts and Guin,[31] if the Reynolds number for flow along the fracture is greater than about 5,000.

Other Models. Barron *et al.*[40] reported the first attempt to model the acid fracturing process using scaled flow experiments. They measured a reaction rate of 15-percent hydrochloric acid during flow between flat, parallel plates of marble in the experimental equipment shown in Fig. 6.10. In these experiments, the fracture was oriented horizontally

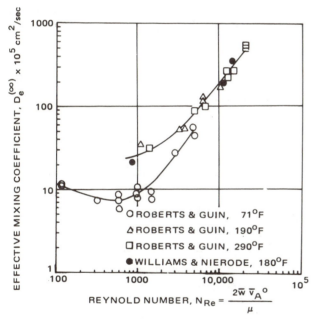

Fig. 6.8—Effective mixing coefficient.[31]

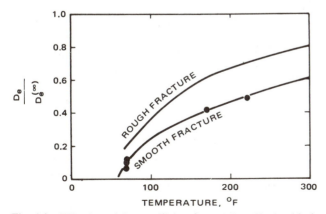

Fig. 6.9—Effective mixing coefficient for reaction of hydrochloric acid and Kasota dolomite.[34]

and temperature was 80°F. Results from these experiments were fitted with an empirical model and a design chart was derived.

Although the study reported that a model for the acid fracturing process must accurately represent the acid flow characteristics, the developed model did not allow accurate prediction of the acid penetration distance. This procedure is inaccurate for the following reasons.

1. All experimental data were taken at room temperature.

2. The fracture model had smooth walls and, therefore, did not accurately represent the fracture geometry expected in a hydraulically induced fracture.

3. The theory used to relate laboratory data to field results did not include the effect of fluid loss or changing flow velocity along the fracture.

4. The experimental model for the fracture was run with the fracture horizontal, thereby minimizing mixing caused by density changes in the acid with reaction.

The design chart proposed by Barron *et al.* is compared with the Nierode-Williams model in Fig. 6.11, showing that the Barron *et al.* model predicts larger penetration distances. These differences can be related to the experimental technique and to procedures used to derive the design chart from the experiments.

The first procedure used to model acid-reaction characteristics was the static reaction rate test.[41-49] In this experiment, a sample of formation rock is contacted by acid and the acid concentration is determined as a function of time. Rock surface area and acid volume are scaled to represent the area-to-volume ratio expected in the fracturing treatment. The time required for acid to react to 10 percent of the initial concentration often is reported as the reaction time for the acid-rock system. Penetration distance is computed by determining the average flow velocity along the fracture in the treatment and then computing the distance acid can travel during its reaction time.

Reaction data reported by Hendrickson *et al.*[50] (Figs. 6.12 and 6.13) are typical of results obtained in static tests. These data show that as the ratio of rock surface area to acid volume is increased, the time required for total acid reaction decreases, and that the reaction time is a function of formation composition (or, acid solubility). To relate these results to acid fracturing treatments, the area-to-volume ratio in the test normally was scaled to represent that expected in the fracture. Therefore, data in Fig. 6.12, where the area-to-volume ratio is in cm^{-1}, would be assumed to represent acid reaction in fractures with a width of 0.78 in. (1:1), 0.1 in. (8:1), and 0.036 in. (22:1).

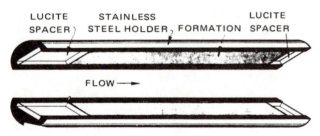

Fig. 6.10—Experimental fracture model, Barron and Wieland.[40]

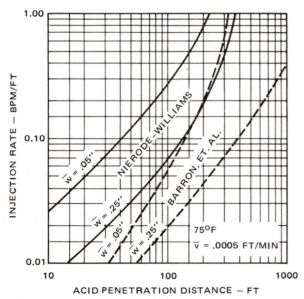

Fig. 6.11—Comparison of acid penetration distance predicted by the Nierode-Williams model with that predicted by the model of Barron and Wieland.[40]

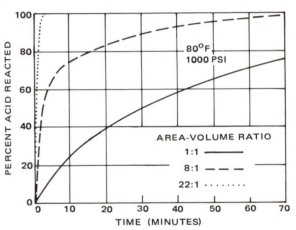

Fig. 6.12—Spending time data for hydrochloric acid and limestone, Hendrickson *et al.*[50]

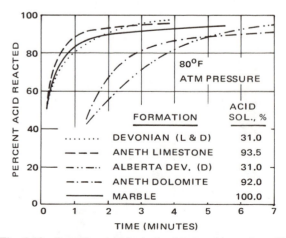

Fig. 6.13—Spending time for hydrochloric acid reaction with formation samples, Hendrickson *et al.*[50]

TABLE 6.3—REACTION TIMES FOR 15-PERCENT HCl[49]

Fracture Width (in.)	Reaction Time (min)		
	80°F	150°F	200°F
0.1	14	7	3
0.24	29	20	12

The primary objection to the static test as a model for acid fracturing treatments is that it does not model mixing induced by acid flow along the rough-walled fracture shown in Fig. 6.7. Williams *et al.*[51] have shown that during the static test, fluid is not truly static, but is circulating because of changes in density associated with the reaction. This fluid motion moves acid to the reactive surface at approximately 10 times the rate expected if the process were controlled by molecular diffusion and can compare with the mixing rate observed in fracturing treatments run with Reynolds numbers of less than 500. Therefore, the static test can give a reasonable representation for acid penetration at low injection rates, but it cannot accurately model acid reaction in most field treatments. *Results from a static test should be assumed to give a qualitative comparison between acid reaction rates in the fracture, but should not be used to predict acid penetration distance.*

Acid penetration distance predicted using the typical reaction-time data from a static reaction rate test given in Table 6.3 is contrasted with the penetration distance predicted by the Nierode-Williams model in Fig. 6.14. This comparison shows that at a fluid-loss velocity, $\bar{v}_N$, of 0.0005 ft/min, the predicted acid penetration distances agree within about 25 percent at injection rates below 0.3 bbl/min/ft. At a fluid-loss velocity of 0.0025 ft/min, they are comparable only at injection rates below 0.15 bbl/min/ft. In some instances, the static test has been analyzed assuming the acid flow velocity is the injection velocity into the fracture rather than the average velocity. If this analysis was used, the static test would not agree at any reasonable injection rate.

6.6 Fracture Conductivity

The conductivity of a fracture created by acid reaction is probably impossible to predict from first principles. Prediction is difficult because the conductivity is a function of the rock strength, heterogeneties present in the rock, the volume and distribution of rock dissolved, and other variables. One estimate of the conductivity can be obtained by assuming the fracture walls are uniformly dissolved, leaving an open channel of width w_a, defined by Eq. 6.36.

$$w_a = \frac{Xit}{2xLh\,(1-\phi)} \,. \qquad \qquad (6.36)$$

In this expression, X is the acid dissolving power (defined in Chapter 3 — values tabulated in Table 3.4), i is the injection rate, and t is the time for which acid is injected. xL and h are the acid penetration distance and fracture height, respectively. The ideal fracture conductivity is, therefore,

$$wk_{fi} = 9.36 \times 10^{13} \left(\frac{w_a}{12}\right)^3, \qquad \qquad (6.37)$$

where w_a is in inches and wk_{fi} is in millidarcy-inches.

This conductivity is usually considerably larger than that observed in laboratory tests, primarily because the effect of closure stress is not considered.

Because of the difficulty in predicting fracture conductivity, the industry was forced either to assume a conductivity believed to be representative or to obtain cores from the formation and measure the conductivity created by acid reaction using techniques similar to those reported by Knox *et al.*[49] and Broaddus and Knox.[41] In a typical experiment to determine fracture flow capacity, a flat, circular piece of formation core is mounted in the experimental equipment shown in Fig. 6.15. Acid is injected through a hole in the center of the core and is allowed to flow radially between the core and a steel plate. After a given reaction time, the core is lowered to contact the steel plate and a closure stress is applied using a hydraulic press. The fracture conductivity then is measured by injecting water into the fracture at a known rate, measuring the pressure drop, and calculating

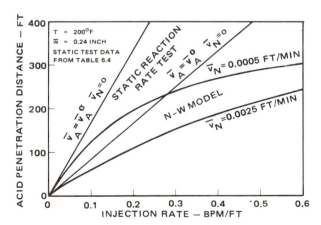

Fig. 6.14—Comparison of acid penetration distance predicted by the Nierode-Williams model with that predicted by a static reaction test.[40]

Fig. 6.15—Experimental equipment used to obtain fracture conductivity.[41]

the apparent flow capacity, using the radial form of Darcy's law. This radial flow test is difficult to interpret because the velocity changes as acid moves radially outward along the cores. For this reason results may not be representative of conductivities expected in a vertical fracture.

Nierode and Kruk[52] have developed a technique for approximating the fracture conductivity without making acidizing studies of formation cores. Their technique allows the actual conductivity to be related to the theoretical ideal conductivity given by Eq. 6.37, the closure stress, σ, and the rock embedment strength, S_{RE},* as shown below.

$$wk_f = C_1 \exp(-C_2\sigma), \quad \ldots \ldots \ldots \ldots \ldots \ldots (6.38)$$

where

$$C_1 = 0.265 (wk_{fi})^{0.822},$$

$$C_2 = (19.9 - 1.3 \ln S_{RE}) \times 10^{-3}$$

$$\text{for } 0 < S_{RE} < 20,000 \text{ psi,}$$

and

$$C_2 = (3.8 - 0.28 \ln S_{RE}) \times 10^{-3}$$

$$\text{for } 20,000 < S_{RE} < 500,000 \text{ psi,}$$

and

$$\sigma = \text{closure stress} = (\text{fracture gradient} \times \text{depth}) - \text{producing bottom-hole pressure.}$$

Eq. 6.38 was obtained by correlating the results of laboratory tests run as follows:

1. A formation core plug was broken in tension to form a rough surface similar to that expected in the fracture.

2. The core was mounted in a triaxial load cell and acid was injected through the fracture. The inlet and outlet concentrations were monitored to allow the volume of rock dissolved to be calculated.

3. Pressure was applied to the load cell, thereby applying

*Rock embedment strength, S_{RE}, is defined as the force required to push a steel ball bearing into the rock surface to a distance equal to the radius of the ball, divided by the projected area of the bearing.

TABLE 6.5—MEASURED ROCK EMBEDMENT STRENGTH OF VARIOUS DRY CARBONATE ROCKS[52]

Formation	Rock Embedment Strength (psi)
Desert Creek B limestone	42,000
San Andres dolomite	50,000 to 175,000
Austin chalk — Buda limestone	20,000
Bloomberg limestone	93,000
Caddo limestone	38,000
Canyon limestone	50,000 to 90,000
Capps limestone	50,000 to 85,000
Cisco limestone	40,000
Edwards limestone	53,000
Indiana limestone	45,000
Novi limestone	106,000
Penn limestone	48,000
Wolfcamp limestone	63,000
Clearfork dolomite	49,000 to 200,000
Greyburg dolomite	75,000 to 145,000
Rodessa Hill laminate	170,000
San Angelo dolomite	100,000 to 160,000

a stress to try to close the fracture. The fracture conductivity was then measured at different stresses.

Results of Nierode and Kruk's tests are given in Table 6.4. Measured embedment strengths for several typical carbonate formations are given in Table 6.5 to use when formation core samples are not available for embedment strength tests.

Nierode and Kruk found that conductivity occurred primarily because of the smoothing of some peaks and valleys on the rock fracture faces. Only limited effects were found to result from rock heterogeneties, apparently caused by the small sample size used in the test. In a few experiments, unusually high conductivity occurred because of a heterogeneity in the core. Since most reservoirs are heterogeneous and acid channeling through a viscous pad fluid to give localized reaction apparently can occur, Eq. 6.38 may underestimate the conductivity actually obtained in the field on occasion. This relation, therefore, should give a conservative estimate of the stimulation that may result from the treatment.

Novotny[53] recently has combined the Nierode conductivity prediction with the model for acid reaction developed by Nierode and Williams[9] to improve the prediction of stimula-

TABLE 6.4—FRACTURE CONDUCTIVITY DATA[52]

Formation	wk_{fi} (md-in.)	S_{RE} (psi)	Conductivity (md-in.) vs Closure Stress (psi) 0	1,000	3,000	5,000	7,000
San Andres dolomite	2.7×10^6	76,000	1.1×10^4	5.3×10^3	1.2×10^3	2.7×10^2	6.0×10^0
San Andres dolomite	5.1×10^9	63,800	1.2×10^6	7.5×10^5	3.0×10^5	1.2×10^5	4.7×10^4
San Andres dolomite	1.9×10^7	62,700	2.1×10^5	9.4×10^4	1.9×10^4	3.7×10^3	7.2×10^2
Canyon limestone	1.3×10^8	88,100	1.3×10^6	7.6×10^5	3.1×10^5	4.8×10^4	6.8×10^3
Canyon limestone	4.6×10^6	30,700	8.0×10^5	3.9×10^5	9.4×10^4	2.3×10^4	5.4×10^3
Canyon limestone	2.7×10^8	46,400	1.6×10^5	6.8×10^5	1.3×10^5	2.3×10^4	4.4×10^3
Cisco limestone	1.2×10^5	67,100	2.5×10^3	1.3×10^3	3.4×10^2	8.8×10^1	2.3×10^1
Cisco limestone	3.0×10^5	14,800	7.0×10^3	3.4×10^3	8.0×10^2	1.9×10^2	4.4×10^1
Cisco limestone	2.0×10^6	25,300	1.4×10^5	6.2×10^4	1.3×10^4	2.7×10^3	5.7×10^2
Capps limestone	3.2×10^5	13,000	9.7×10^3	4.2×10^3	7.6×10^2	1.4×10^2	2.5×10^1
Capps limestone	2.9×10^5	30,100	1.8×10^4	6.8×10^3	1.4×10^2	1.3×10^2	1.8×10^1
Indiana limestone	4.5×10^6	22,700	1.5×10^5	1.5×10^5	1.5×10^3	1.5×10^3	1.5×10^2
Indiana limestone	2.8×10^7	21,500	7.9×10^5	3.0×10^5	4.3×10^4	6.3×10^3	9.0×10^2
Indiana limestone	3.1×10^8	14,300	7.4×10^6	2.0×10^6	1.4×10^5	1.0×10^4	7.0×10^2
Austin chalk	3.9×10^6	11,100	5.6×10^4	1.6×10^3	1.3×10^0	—	—
Austin chalk	2.4×10^6	5,600	3.9×10^4	1.2×10^3	1.2×10^0	—	—
Austin chalk	4.8×10^5	13,200	1.0×10^4	1.7×10^3	4.9×10^1	1.4×10^0	—
Clearfork dolomite	3.6×10^4	35,000	3.4×10^3	1.7×10^3	4.1×10^2	1.0×10^2	2.4×10^1
Clearfork dolomite	3.3×10^4	11,800	9.3×10^3	1.6×10^3	4.5×10^1	1.3×10^0	—
Greyburg dolomite	8.3×10^6	14,400	2.5×10^5	4.0×10^4	1.0×10^3	2.5×10^1	—
Greyburg dolomite	3.9×10^6	12,200	2.1×10^5	7.9×10^4	1.0×10^4	1.5×10^3	2.0×10^2
Greyburg dolomite	3.2×10^6	16,600	8.0×10^4	1.5×10^4	2.8×10^2	1.6×10^1	—
San Andres dolomite	1.0×10^6	46,500	8.3×10^4	4.0×10^4	9.5×10^3	2.2×10^3	5.2×10^2
San Andres dolomite	2.4×10^6	76,500	1.9×10^4	6.8×10^3	8.5×10^2	1.0×10^2	1.3×10^1
San Andres dolomite	3.4×10^6	17,300	9.4×10^3	2.8×10^3	2.5×10^2	2.3×10^1	—

tion from acidization (discussed in Section 6.5). In his model, he keeps track of the volume of acid reacted at each position along the fracture. This allows prediction of the theoretical fracture width as a function of position. A typical fracture width prediction is shown in Fig. 6.16 for no closure stress and for closure stresses of 1,000 and 10,000 psi. Note that the width decreases rapidly with distance from the wellbore and that the fracture width (and thereby conductivity) decreases with closure stress.

To validate his model, Novotny reanalyzed data from treatment of the 27 oil and gas wells considered before by Nierode et al.,[34] and analyzed 16 additional wells. A comparison of predicted with observed stimulation ratios for these cases is given in Fig. 6.17. For comparison, a plot with prediction assuming infinite fracture conductivity, as proposed by Nierode et al., is given as Fig. 6.18. The assumption of finite conductivity gave a significant statistical improvement in the comparison with field results.

References

1. Geertsma, J. and de Klerk, F.: "A Rapid Method of Predicting Width and Extent of Hydraulically Induced Fractures," *J. Pet. Tech.* (Dec. 1969) 1571-1581.

2. Baron, G.: "Fracturation hydraulique; bases théoriques, études de laboratoire, essais sur champ," *Proc.*, Seventh World Pet. Cong., Mexico City (1967) 371-393.

3. Daneshy, A. A.: "On the Design of Vertical Fractures," *J. Pet. Tech.* (Jan. 1973) 83-97.

4. Kiel, O. M.: "A New Hydraulic Fracturing Process," *J. Pet. Tech.* (Jan. 1970) 89-96.

5. Nordgren, R. P.: "Propagation of a Vertical Hydraulic Fracture," *Soc. Pet. Eng. J.* (Aug. 1972) 306-314.

6. Perkins, T. K. and Kern, L. R.: "Width of Hydraulic Fracture," *J. Pet. Tech.* (Sept. 1961) 937-949.

7. Perkins, T. K. and Krech, W. W.: "The Energy Balance Concept of Hydraulic Fracturing," *Soc. Pet. Eng. J.* (March 1968) 1-12.

8. Sinclair, A. R.: "Heat Transfer Effects in Deep Well Fracturing," *J. Pet. Tech.* (Dec. 1971) 1484-1492.

9. Williams, B. B. and Nierode, D. E.: "Design of Acid Fracturing Treatments," *J. Pet. Tech.* (July 1972) 849-859.

10. Khristianovich, S. A. and Zheltov, J. P.: "Formation of Vertical Fractures by Means of a Highly Viscous Liquid," *Proc.*, Fourth World Pet. Cong., Rome (1955) 579-586.

11. Zheltov, J. P. and Khristianovich, S. A.: "The Hydraulic Fracturing of an Oil-Producing Formation," *Otdel Tekh Nauk*, Izvest. Akad. Nauk USSR (May 1955) 3-41.

12. Barenblatt, G. I.: "The Mathematical Theory of Equilibrium Cracks and Brittle Fracture," *Advances in Applied Mechanics*, Academic Press, New York (1962) 55-129.

13. Howard, G. C. and Fast, C. R.: "Optimum Fluid Characteristics for Fracture Extension," *Drill. and Prod. Prac.*, API (1957) 261-270.

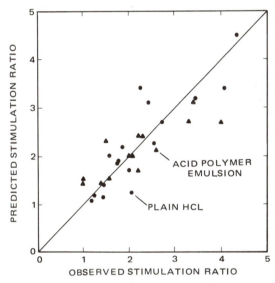

Fig. 6.17—Comparison of predicted with observed stimulation ratios, assuming *finite* acidized fracture conductivity.[53]

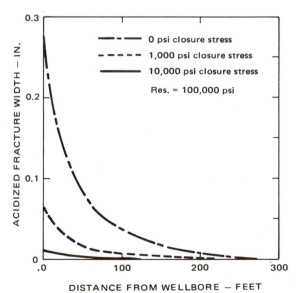

Fig. 6.16—Acidizing fracture width as a function of the distance from the wellbore and closure stress.[53]

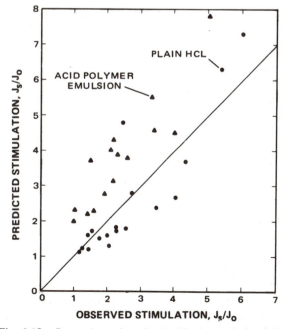

Fig. 6.18—Comparison of predicted with observed stimulation ratios, assuming *infinite* acidized fracture conductivity.[53]

14. Howard, G. C. and Fast, C. R.: *Hydraulic Fracturing,* Monograph Series, Society of Petroleum Engineers of AIME, Dallas (1970).

15. Williams, B. B.: "Fluid Loss From Hydraulically Induced Fractures," *J. Pet. Tech.* (July 1970) 882-888.

16. Whan, G. A. and Rothfus, R. R.: "Characteristics of Transition Flow Between Parallel Plates," *AIChE Jour.* (June 1959) 204-208.

17. Eickmeier, J. R. and Ramey, H. J., Jr.: "Wellbore Temperature and Heat Losses During Production or Injection Operations," paper presented at the CIM 21st Annual Technical Meeting, Calgary, Alta., May 6-8, 1970.

18. Moss, J. T. and White, P. D.: "How to Calculate Temperature Profiles in Water Injection Wells," *Oil and Gas J.* (March 9, 1959) 174-178.

19. Ramey, H. J., Jr.: "Wellbore Heat Transmission," *J. Pet. Tech.* (April 1962) 427-435.

20. Squier, D. P., Smith, D. D., and Dougherty, E. L.: "Calculated Temperature Behavior of Hot Water Injection Wells," *J. Pet. Tech.* (April 1962) 436-440.

21. Leutwyler, K.: "Casing Temperature Studies in Steam Injection Wells," *J. Pet. Tech.* (Sept. 1966) 1157-1162; *Trans.,* AIME, **237**.

22. Claycomb, J. R. and Schweppe, J. L.: "A Numerical Method for Determining Circulation and Injection Temperatures," paper 67-PET-3 presented at the ASME Petroleum Mechanical Engineering Conference, Philadelphia, Sept. 17-20, 1967.

23. Raymond, L. R.: "Temperature Distribution in a Circulating Drilling Fluid," *J. Pet. Tech.* (March 1969) 333-342.

24. Keller, H. H., Couch, E. J., and Berry, P. M.: "Temperature Distribution in Circulating Mud Columns," *Soc. Pet. Eng. J.* (Feb. 1973) 23-30.

25. Sump, G. D. and Williams, B. B.: "Prediction of Wellbore Temperatures During Mud Circulation and Cementing Operations," *Trans.,* ASME (1973) **95B**, 1083-1092.

26. Wheeler, J. A.: "Analytical Calculations of Heat Transfer from Fractures," paper SPE 2494 presented at the SPE-AIME Improved Oil Recovery Conference, Tulsa, Okla., March 1969.

27. Sinclair, A. R.: "Heat Transfer Effects in Deep Well Fracturing," *J. Pet. Tech.* (Dec. 1971) 1484-1492.

28. Dysart, G. R. and Whitsitt, N. F.: "Fluid Temperatures in Fractures," paper SPE 1902 presented at the SPE-AIME 42nd Annual Fall Meeting, Houston, Oct. 1-4, 1967.

29. Hill, W. L. and Wahl, H. A.: "The Effect of Formation Temperature on Hydraulic Fracture Design," paper 851-42-1 presented at the API Spring Meeting, Mid Continent Div., Amarillo, Tex., April 1968.

30. Whitsitt, N. F. and Dysart, G. R.: "The Effect of Temperature on Stimulation Design," *J. Pet. Tech.* (April 1970) 493-502; *Trans.,* AIME, **249**.

31. Roberts, L. D. and Guin, J. A.: "The Effect of Surface Kinetics in Fracture Acidizing," *Soc. Pet. Eng. J.* (Aug. 1974) 385-395; *Trans.,* AIME, **257**.

32. Roberts, L. D. and Guin, J. A.: "A New Method for Predicting Acid Penetration Distance," *Soc. Pet. Eng. J.* (Aug. 1975) 277-286.

33. Nierode, D. E. and Williams, B. B.: "Characteristics of Acid Reaction in Limestone Formations," *Soc. Pet. Eng. J.* (Dec. 1971) 406-418.

34. Nierode, D. E., Williams, B. B., and Bombardieri, C. C.: "Prediction of Stimulation From Acid Fracturing Treatments," *J. Cdn. Pet. Tech.* (Oct.-Dec. 1972) 31-41.

35. Whitsitt, N. F., Harrington, L. J., and Hannah, B.: *A New Approach to Deep Well Acid Stimulation Design,* The Western Co., Fort Worth, Tex. (June 10, 1970).

36. van Domselaar, H. R., Schols, R. S., and Visser, W.: "An Analysis of the Acidizing Process in Acid Fracturing," *Soc. Pet. Eng. J.* (Aug. 1973) 239-250.

37. Bird, R. B., Stewart, W. E., and Lightfoot, E. N.: *Transport Phenomena,* John Wiley & Sons, Inc., New York (1960).

38. Gill, W. N.: "Convective Diffusion in Laminar and Turbulent Reverse Osmosis Systems," *Surface and Colloid Science,* John Wiley & Sons, Inc., New York (1971) **4**, 263.

39. Terrill, R. M.: "Heat Transfer in Laminar Flow Between Parallel Porous Plate," *Int. J. Heat Mass Transfer* (1965) **8**, 1491-1497.

40. Barron, A. N., Hendrickson, A. R., and Wieland, D. R.: "The Effect of Flow on Acid Reactivity in a Carbonate Fracture," *J. Pet. Tech.* (April 1962) 409-415; *Trans.,* AIME, **225**.

41. Broaddus, G. C. and Knox, J. A.: "Influence of Acid Type and Quantity in Limestone Etching," paper 851-39-I presented at the API Mid-Continent Meeting, Wichita, Kans., March 31-April 2, 1965.

42. Chamberlain, L. C. and Boyer, R. F.: "Acid Solvents for Oil Wells," *Ind. and Eng. Chem.* (1939) **31**, 400.

43. Dill, W. R.: "Reaction Times of Hydrochloric-Acetic Acid Solution on Limestone," paper presented at the ACS 16th Southwest Regional Meeting, Oklahoma City, Dec. 1-3, 1960.

44. Dunlop, P.: "Mechanisms of the HCl-$CaCO_3$ Reaction," paper presented at the ACS 17th Southwest Regional Meeting, Oklahoma City, Dec. 1961.

45. Harrison, N. W.: "A Study of Stimulation Results of Ellenberger Gas Wells in the Delaware-Val Verde Basins," *J. Pet. Tech.* (Aug. 1967) 1017-1021.

46. van Poollen, H. K.: "How Acids Behave in Solution," *Oil and Gas J.* (Sept. 25, 1967) **65**, 100-102.

47. van Poollen, H. K. and Jorgon, J. R.: "How Conditions Affect Reaction Rate of Well-Treating Acids," *Oil and Gas J.* (Oct. 21, 1968) **66**, 84-91.

48. Staudt, J. G. and Love, W. W.: "Sustained Action Acid," *World Pet.* (May 1938) 78-80.

49. Knox, J. A., Pollock, R. W., and Beechcroft, W. H.: "The Chemical Retardation of Acid and How It Can Be Utilized," *J. Cdn. Pet. Tech.* (Jan.-March 1965) 5-12.

50. Hendrickson, A. R., Rosene, R. B., and Wieland, D. R.: "Acid Reaction Parameters and Reservoir Characteristics Used in the Design of Acid Treatment," paper presented at the ACS 137th Meeting, Cleveland, Ohio, April 5-14, 1960.

51. Williams, B. B., Gidley, J. L., Guin, J. A., and Schechter, R. S.: "Characterization of Liquid/Solid Reactions — The Hydrochloric Acid/Calcium Carbonate Reaction," *I. and E.C. Fund.* (Nov. 1970) **9**, 589-596.

52. Nierode, D. E. and Kruk, K. F.: "An Evaluation of Acid Fluid Loss Additives, Retarded Acids, and Acidized Fracture Conductivity," paper SPE 4549 presented at the SPE-AIME 48th Annual Fall Meeting, Las Vegas, Sept. 30-Oct. 3, 1973.

53. Novotny, E. J.: "Prediction of Stimulation From Acid Fracturing Treatments Using Finite Fracture Conductivity," *J. Pet. Tech.* (Sept. 1977) 1186-1194; *Trans.,* AIME, **262**.

Chapter 7

Acid Fracturing Treatment Design

7.1 Introduction

This chapter defines one possible procedure for designing an acid fracturing treatment after the candidate well is selected. Normally, the treatment to be used is specified only after designing several possible treatments, estimating the productivity increase that can be expected and the life of the productivity increase for each treatment, and then conducting a detailed economic evaluation using procedures discussed in Chapter 12. The treatment design (pad-fluid type and volume, acid concentration and volume, injection rate, etc.) that maximizes the risk-weighted economic return normally would then be selected.

The selection of a candidate well is an important step in this process. The first well for treatment in a reservoir normally should be the one expected to give the greatest productivity increase with minimal down-side risk. For this reason, a reservoir study, including well tests, should be conducted when possible to determine a priority list for stimulation. This list of possible candidates then should be investigated for possible production problems — for example, old casing or tubing, a poor cement job between the zone to be treated and an adjacent aquifer, or a water column within the zone to be treated. Wells that have a high probability of trouble should be downgraded as potential stimulation candidates.

The design of an acid fracturing treatment to stimulate production involves the following five steps.

1. Determine formation rock and fluid properties.

2. Select variable parameters, including the fracturing fluid to be used as a pad, injection rate, etc.

3. Predict the fracture geometry and the acid penetration distance for the fracturing fluid and acid of interest.

4. Predict the fracture conductivity and expected stimulation ratio for pad and acid volumes of interest.

5. Select the most economic treatment.

Each step is discussed and illustrated in the remainder of this chapter. A conceptual discussion of acid fracturing is included in Chapter 5. Models used in these example calculations are discussed in Chapter 6. We believe models chosen for use here are among the more accurate available and yet are simple enough to allow calculations with only a slide rule or calculator. However, as noted in Chapter 6, other models can be used.

7.2 Determination of Formation Matrix and Fluid Properties

To predict the geometry of a fracture created in an acid fracturing treatment, it is necessary to have accurate reservoir data. The selection of good input data is often the most important part of a successful treatment design. This point cannot be overemphasized.

Matrix and fluid properties generally are assigned average values that remain fixed for all calculations in that particular reservoir. A list of fixed parameters required to design a fracturing treatment is given in Table 7.1. Each parameter is discussed in this section.

Formation Thickness

- Gross (h_g) — Total expected vertical fracture height, ft.
- Net (h_n) — Portion of the vertical fracture height that will accept fluid during the fracturing treatment, ft.

Fracture design calculations are very sensitive to variations in these fracture heights; therefore, the expected vertical extent of the fracture should be estimated as accurately as possible.

It is important to recognize that we can measure fracture height only at the wellbore. There is no proven technique for estimating fracture height long distances from the wellbore; therefore, fracture height is seldom known with certainty. Field data indicate that in some areas with competent, geologically young shales, a 10- to 20-ft-thick shale body

TABLE 7.1—FORMATION PROPERTIES REQUIRED FOR ACID TREATMENT DESIGN

Formation thickness, ft
  Gross (fracture height)
  Net (permeable)
Permeability, md
Porosity, fraction
Depth, ft
Fracturing gradient, psi/ft
Poisson's ratio
Sonic travel time, μsec/ft
Temperature, °F
  Formation
  Fluid injection
Reservoir pressure, psi
Reservoir fluid properties
  Viscosity, cp
  Compressibility, psi^{-1}
  Density, lb/cu ft

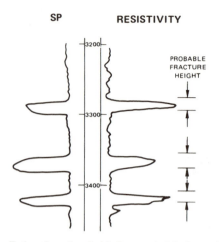

Fig. 7.1—Estimation of probable fracture height for multiple strata separated by thick barriers.

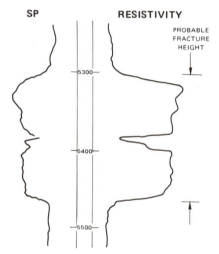

Fig. 7.2—Estimation of probable fracture height for multiple strata separated by thin barriers.

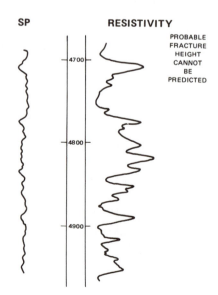

Fig. 7.3—Massive strata with no thick barriers (using temperature logs or other well data).

often will terminate vertical fracture growth when a low-viscosity gelled fluid or acid is used, but that a 20- to 30-ft shale may be required to terminate growth when a highly viscous fluid is used. When shales are calcareous, or very hard and geologically old, they often do not readily terminate fracture growth. To determine fracture penetration into these shales, temperature surveys should be run following the first treatments in the field. In any location, a conservative treatment design will result when the fracture is assumed to have the maximum expected vertical height.

Figs. 7.1 through 7.3 are example logs (SP on left and resistivity on right) that illustrate the best estimate for fracture height when only log data are available.

Fig. 7.1 is an example in which a fracture is likely to remain within the shale boundaries indicated. For treatment of one of these three zones, the gross fracture height, h_g, would be estimated as the height of the zone, plus 10 to 20 ft for a conventional treatment or plus 30 ft in a treatment where a viscous fluid is used, to account for penetration into the bounding formations. Net height, h_n, in each case is the permeable formation thickness. (A treatment in which a very viscous fluid is used could possibly fracture both of the lower sections simultaneously.)

Fig. 7.2 is an example in which a fracture is expected to extend through the entire section from 5,300 to 5,470 ft. The break at 5,390 ft might terminate fracture growth if a low-viscosity fluid is used, but this is not certain and a definite conclusion requires prior experience. In this example, a gross height of 190 ft and net of 120 ft would be selected.

Fig. 7.3 is an example for which there is no basis for choosing the fracture height from a log. To estimate fracture height in these formations, we must resort to prior data, usually temperature logs from previously fractured wells. (In some cases the sand has been tagged with a radioactive mineral and height determined from a radiation survey.[26])

In massive formations, the fracture will initiate at the perforations and grow symmetrically outward until it reaches a boundary that will restrain further vertical growth. For this reason, the expected fracture height is sometimes assigned as a function of the total fluid injected (total fracture volume) or the injection rate. This technique for estimating vertical fracture height may be accurate when used to interpolate within the range of treating conditions for which the correlation was derived. *Extreme caution* is necessary, however, when attempting to extrapolate to new formations, different geographical locations, different fracturing fluids, or larger-volume fracture treatments. Correlations relating fracture height to treatment parameters are normally derived for a given reservoir and often can be obtained from service companies.

Formation Permeability

The average permeability of the formation to the fracturing fluid (not air) is required. Generally, permeability data from a pressure buildup or injectivity falloff test are preferred. If well test data are not available, core data or the best approximation from undamaged initial productivity should be used. These base permeabilities should be altered to account for relative permeability effects, using laboratory-determined relative permeability curves, when available. When relative permeability data are not available, a rule of thumb we often use is to divide the absolute permeability by 1.5 for oil-based fracturing fluids and by 5 for water-based fracturing fluids.

Formation Porosity

Use the average porosity of the reservoir determined from log or core analysis; porosity normally is not a critical parameter in treatment design.

Formation Depth

Use the distance from ground level to the middle of the formation, expressed in feet.

Formation Fracturing Gradient

Use the pressure, expressed as the gradient (psi per foot of depth), required to hold open the fracture just as the fracture walls are about to close. This is not the pressure required to initiate a fracture, often called the breakdown pressure. The breakdown pressure will normally exceed the fracture propagation pressure as measured by the fracturing gradient defined below.

The fracture gradient is estimated by adding the surface pressure observed instantaneously after shut-in of the fracturing pumps to the hydrostatic head of the fluid in the wellbore, and dividing by formation depth.

$$g_f = \frac{\text{instantaneous shut-in pressure} + \text{hydrostatic head}}{\text{depth}}$$

$$\dotfill (7.1)$$

The fracture gradient for a reservoir is not constant, but changes as reservoir pressure is changed. Two example surface-pressure records for fracture treatments in the same reservoir are presented in Fig. 7.4 to illustrate this effect. The first treatment was in a newly developed field with an initial reservoir pressure of 2,000 psi; the fracture gradient was 0.7 psi/ft. At the time of the second treatment, reservoir pressure had been depleted to 1,000 psi; the fracture gradient was 0.6 psi/ft. From this example, it is apparent that a method is needed to predict changes in fracture gradient with reservoir pressure, to estimate the fracture gradient for deep, high-pressure reservoirs, and to predict the gradient for newly developed fields.

A simple, approximate method that can be used to estimate the fracture gradient is to assume it is proportional to the overburden and reservoir pressure gradients:

$$g_f \tilde{=} \alpha + (\text{overburden gradient} - \alpha)\frac{\text{reservoir pressure}}{\text{depth}},$$

$$\dotfill (7.2)$$

where α = constant (0.33 to 0.5) and the overburden gradient is about 1.0 psi/ft at depths less than 10,000 ft and 1.0 to 1.2 psi/ft at depths greater than 10,000 ft. To extrapolate to a reduced reservoir pressure from a fracture gradient obtained at initial reservoir pressure, substitute the old fracture gradient and reservoir pressure into Eq. 7.2 and calculate α. Then, for the new reservoir pressure and the computed value of α, estimate the revised fracturing gradient. More rigorous

theories for predicting the fracture gradient as a function of changes in overburden stress, pore pressure, rock tensile strength, etc. exist. We seldom have sufficient data to use these theories, however, and the simple relationship given in Eq. 7.2, although not exact, is usually adequate.

In a new field, the fracture gradient can be approximated using Eq. 7.2 with $\alpha = 0.5$. *Remember, this is an approximate equation* and that the proper value of α should be verified from field data.

Poisson's Ratio

The fracture geometry predicted for a given formation will vary only slightly as Poisson's ratio for the rock varies. Therefore, it is sufficient to select values for Poisson's ratio based upon the general rock type. Typical values are listed below.

Rock Type	Poisson's Ratio
Hard carbonate	0.25
Medium-hard carbonate	0.27
Soft carbonate	0.30

Mean Sonic Travel Time

The mean sonic travel time can be used to calculate the modulus of elasticity of the formation (Young's modulus). It is best to use travel times from sonic logs taken in the well to be treated. Since Young's modulus, which is an important parameter in the prediction of fracture geometry, is normally evaluated from the sonic travel time, the travel time must be measured as accurately as possible. If in doubt, select the

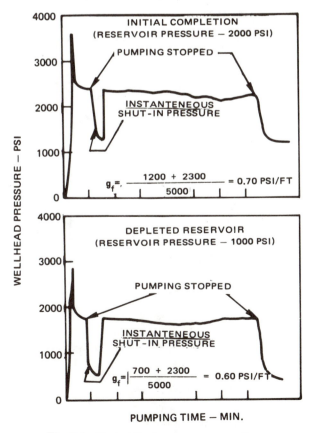

Fig. 7.4—Typical fracture pressure recordings.

minimum expected travel time to assure that any error in fracture geometry prediction is conservative — that is, that the observed fracture width will be larger than the calculated width. Young's modulus for the formation can be estimated from the measured sonic travel time using Fig. 7.5.[1] A more accurate value can be calculated using Eq. 7.3.[1]

$$E = 2.16 \times 10^8 \frac{[\rho_{ma}(1-\phi) + \rho_{fl}\phi](1-2\nu)(1+\nu)}{(1-\nu)t_s^2},$$

$$\dots \dots \dots \dots \dots (7.3)$$

where

t_s = sonic travel time, μsec/ft
ϕ = porosity
ν = Poisson's ratio
ρ_{ma} = density of the formation matrix, lb/cu ft
ρ_{fl} = density of the formation fluids, lb/cu ft
E = Young's modulus, psi.

Formation Temperature

This is the temperature of the formation to be treated. The temperature can be obtained most accurately from a temperature survey in the well of interest.

Fluid Injection Temperature

The fluid injection temperature is defined as the temperature of the fluid as it enters the fracture. This temperature can be estimated using techniques described in Chapter 6, Section 6.4.

Reservoir Pressure

The average reservoir pressure should be taken from the most recent reservoir data.

Formation Fluid Viscosity

This is the viscosity of the reservoir fluids at reservoir conditions. The variation of viscosity with pressure and oil gravity is normally available for the reservoir oil. The viscosity of the fluid (water, oil, or gas) expected to be mobile in the reservoir should be used in the calculation.

Figs. 7.6 through 7.8 present correlations for oil, water, and gas viscosities to be used to estimate the viscosity of the mobile formation fluids.[2-6]

Formation Fluid Compressibility

The isothermal coefficient of compressibility usually is determined from PVT data. If this information is not available, the compressibility can be estimated as follows.

$$K_{fl} = S_o(K_o) + S_w(K_w) + S_g(K_g), \dots (7.4)$$

where

K_{fl} = isothermal coefficient of compressibility of reservoir fluids, psi^{-1}
S_o, S_w, S_g = oil, water and gas saturation fractions
K_o, K_w, K_g = isothermal coefficient of compressibility of reservoir oil, water, and gas, psi^{-1}.

To estimate the oil compressibility, calculate the pseudo-reduced pressure and temperature by dividing the pressure and temperature by p_{pc} and T_{pc}, the pseudocritical pressure and temperature of the oil. p_{pc} and T_{pc} can be evaluated from Figs. 7.9 and 7.10 given the bubble-point pressure and oil specific gravity.

$$p_{pr} = \frac{p}{p_{pc}}, T_{pr} = \frac{T}{T_{pc}} \dots (7.5)$$

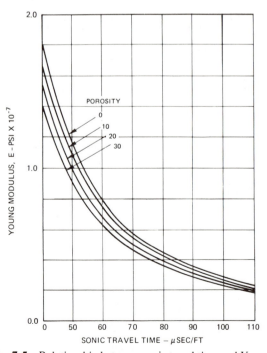

Fig. 7.5—Relationship between sonic travel time and Young's modulus of carbonate formations.

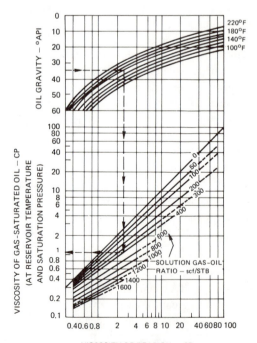

Fig. 7.6—Gas-saturated oil viscosity at reservoir temperature and pressure.[2,3]

For the computed pseudo-reduced pressure and temperature, a pseudo-reduced compressibility, K_{pr}, can be obtained from Fig. 7.11. The oil compressibility is then calculated as

$$K_o = \frac{K_{pr}}{p_{pc}} \quad \dots\dots\dots\dots\dots\dots\dots\dots\dots\dots (7.6)$$

Water compressibility, K_w, can be read from Fig. 7.7 for

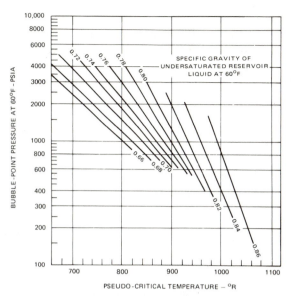

Fig. 7.9—Variation of pseudocritical temperature with specific gravity and oil bubble point.

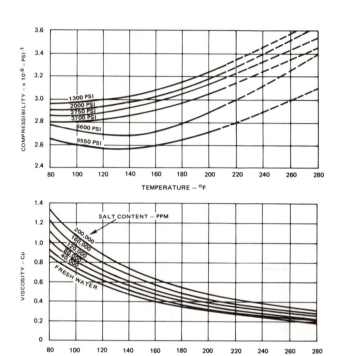

Fig. 7.7—Brine viscosity and compressibility as a function of temperature and salinity.[4,5]

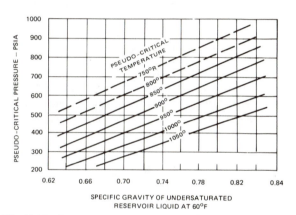

Fig. 7.10—Pseudocritical pressure as a function of oil specific gravity and pseudocritical temperature.

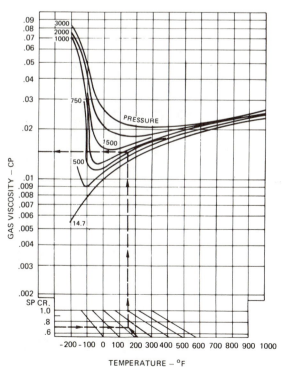

Fig. 7.8—Gas viscosity as a function of specific gravity, temperature, and pressure.[6]

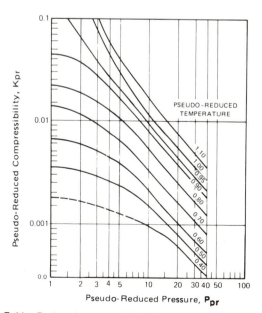

Fig. 7.11—Reduced compressibility for undersaturated reservoir oils.

the reservoir pressure and temperature.

Gas compressibility, K_g, is approximately $1/p$, where p is the reservoir pressure. To calculate a more exact value for K_g, use the following procedure.

- Calculate the pseudo-reduced temperature and pressure, T_{pr} and p_{pr}.

$$T_{pr} = \frac{T}{T_{pc}} \; , \; p_{pr} = \frac{p}{p_{pc}} \; , \; \dots\dots\dots\dots\dots (7.7)$$

where T_{pc} and p_{pc}, the pseudo-critical temperature and pressure for the gas, can be determined from Fig. 7.12.[8]

- From Figs. 7.13A and 7.13B evaluate the pseudo-reduced compressibility, K_{pr}, then calculate K_g as follows.[7]

$$K_g = \frac{K_{pr}}{p_{pc}} \; , \; \dots\dots\dots\dots\dots\dots\dots (7.8)$$

Gas compressibility must be included if the reservoir pressure is at or near the bubble-point pressure. When appreciable gas is expected, the over-all coefficient of compressibility approaches the gas compressibility.

Formation Fluid Density

The average formation fluid density is in pound-mass per cubic foot.

7.3   Selection of Variable Design Parameters

The variable design parameters in an acid fracturing treatment, those over which we have control, include (1) the type and viscosity of the pad fluid, (2) acid concentration and additives to be used, (3) injection rate for the pad fluid and acid, and (4) pad fluid and acid volume.

Pad Fluid Type and Viscosity

The ideal fracturing fluid to be used as a pad preceding the acid should have a number of specific properties. Ideally, the fluid (1) should have a high friction pressure drop while flowing along the fracture to allow creation of a wide fracture; (2) must allow adequate fluid-loss control to create a long and wide fracture; (3) must not react with acid to form precipitates or other material that might restrict well productivity; (4) should pump with a low tubular pressure drop; (5) must be removed easily from the formation without causing permeability damage to the formation; (6) should have low cost; and (7) should be safe to use and easy to handle.

Obviously, no single fluid can meet all these requirements at all times; therefore, there is no single ideal fracturing fluid for use as a pad in acid fracturing treatments.

Both oil and water are used to prepare pad fluids. Because there is obviously a large number of fluids and additives, the systems most often used will be reviewed. Detailed fluid properties can be obtained from the service companies that market these fluids.[9-12] Procedures for obtaining accurate

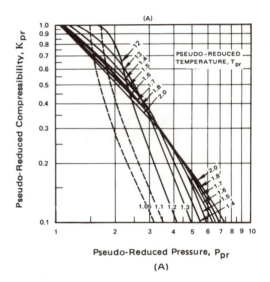

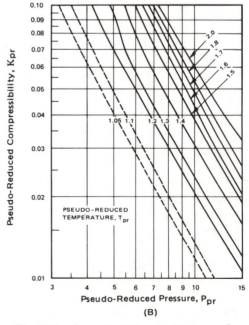

Fig. 7.13—Compressibility of natural gases.[7]

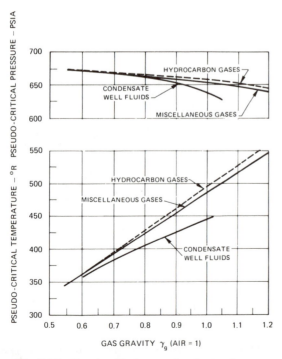

Fig. 7.12—Pseudocritical properties of natural gases.

data are included in a new revison of APIRP 42.

Water-Based Pad Fluids — Under some circumstances, water without gelling agent can be a satisfactory pad fluid. Water, when used, can be without a fluid-loss additive or with an additive that is prepared by coating an inert particle with a gelatinous material. Because water has a low viscosity, it normally cannot create a wide fracture. The main advantage of water as a fracturing fluid is its low cost. Water alone is sometimes used in very hot formations to precool well tubulars, thereby minimizing corrosion and increasing acid penetration distance.

Low-viscosity gelled waters prepared with guar are the most common fracturing fluids used to precede acid. These fluids normally contain silica flour as a fluid-loss additive; however, commercially available inert particles coated with a guar-like material also are used in high-temperature formations. These fluids are limited because of their low viscosity that degrades rapidly with temperature, especially in the presence of acid. Because of this low viscosity, these guar-based fluids normally create a narrow fracture.

High-viscosity gelled waters developed to create wide fractures can be effective pad fluids for acid fracturing. The available high-viscosity gelled waters can be classified as fluids for use at temperatures below about 200°F (prepared using guar or cellulose polymers) and fluids intended for use at temperatures greater than 200°F (prepared using synthetic polymers).

The typical high-viscosity fluids prepared from a guar base are made by dissolving 40 to 80 lb of guar per 1,000 gal of water and then adding a chemical that reacts to link the guar molecules. Because of the gel structure formed after crosslinking, these fluids are very viscous. The fluid viscosity will vary with polymer concentration, temperature, and shear rate.

Fluids for high-temperature application have been developed from commercially available synthetic water-soluble polymers, including derivatives of the cellulose or polyacrylamide types. In some instances, *polyacrylamide polymers* can be crosslinked by acid reaction products in a low pH environment, resulting in the formation of a rubber-like material that can limit well productivity after treatment. Care must be taken to assure this possibility does not exist before using a polyacrylamide-based fluid.

Cellulose-based fluids can normally be used in acid fracturing applications at virtually any reservoir temperature because the gel is readily broken by contact with acid. The use of these fluids has been limited by their high cost relative to many other fracturing fluids.

Viscous emulsions prepared with oil as the internal phase and fresh water, brine, or acid as the external phase[13, 14] also can be used as a pad fluid. This fluid has a high viscosity because the oil content of the emulsion is high (normally 60 to 70 percent) and the external aqueous phase contains a polymer. The fluid is stable if properly prepared and will break when contacted with acid or divalent ions. The prime advantage of the fluid is the high viscosity formed at a low cost.

Oil-Based Pad Fluids — Viscous oils can be used effectively as a pad before acid if they are properly selected for the specific formation temperature and permeability, although they have not been used widely in this application. Viscosity should be less than about 200 cp in formations with a permeability below about 10 md, and should be less than 50 cp at reservoir temperature if the permeability is less than 1 md.

Most petroleum service companies have developed gelling agents for low-viscosity oils; these gelling agents increase oil viscosity without increasing the tubular friction pressure.[10, 12, 15] The gelled oils pump at a relatively low friction pressure, have a moderately high viscosity during flow along the fracture, and break when contacted with formation oils or a chemical breaker. Since these fluids are often very sensitive to acid, tests should be conducted to assure that they do not break down immediately on contact with acid and that damaging precipitates do not form.

Acid Concentration and Additives — Acid concentration is an important parameter in treatment design. Often, high-strength acid (28 percent) is preferred over 15-percent HCl because it has an increased dissolving power and evolves a significant volume of CO_2 that can help to speed cleanup of the well following stimulation. Also, the spent acid is more viscous than the 15-percent HCl, thereby helping to reduce the rate of fluid loss. A more complete discussion of acid types, their limitations, and their advantages is given in Chapter 3.

Additives often used in acidization are discussed in Chapter 11. The most important additives are usually the corrosion inhibitor and acid fluid-loss additive. The corrosion inhibitor must be selected to control corrosion of the acid used with well tubulars at the temperature expected in the tubing. When possible, an additive should be selected to allow corrosion control of reservoir temperature. The fluid-loss additive is often critical to treatment results, as discussed in Chapter 5. *Seldom can stimulation be maximized without an effective acid fluid-loss additive.* Both additives should be selected to give minimum damage following the acid treatment.

Injection Rate

To assure that the selected treatment is the most economic one, several different injection rates should be considered in the treatment design procedure. It is particularly important to recognize that increasing the injection rate does not always improve treatment results, but that it normally increases pumping costs.

The maximum injection rate for the pad fluid or acid can be estimated by computing the maximum allowable friction-pressure gradient, g_{pf}, and then reading the maximum pump rate from friction-pressure curves for the fluid of interest.*

g_{pf} = (maximum allowable surface pressure + hydrostatic head − fracture gradient × depth) ÷ depth

$$\dotfill (7.9)$$

---

*Friction-pressure data can be obtained on request from the petroleum service companies that conduct fracturing and acidizing treatments.

Example Calculation

Consider, for example, that the allowable friction-pressure gradient is desired for a treatment where the following conditions apply: (1) maximum allowable surface pressure = 5,000 psi; (2) fluid specific gravity = 1.2; (3) fracture gradient = 0.7 psi/ft; and (4) formation depth = 5,000 ft.

Substituting in Eq. 7.9, we find

$$g_{pf} = \frac{5{,}000 \text{ psi} + (1.2)(0.43 \text{ psi/ft}) (5{,}000 \text{ ft})}{5{,}000 \text{ ft}}$$

$$\frac{- 0.7 \text{ psi/ft} (5{,}000 \text{ ft})}{5{,}000 \text{ ft}}$$

$$g_{pf} = 0.82 \;\; \frac{\text{psi}}{\text{ft}}$$

As noted, the maximum allowable pump rate for fluids of interest can be obtained from friction-pressure charts available from the oilfield service companies, given this allowable friction-pressure gradient.

7.4 Calculation of Fracture Geometry and Acid Penetration Distance

The dimensions of a fracture created by a pad fluid, if used, or the acid, if a pad is not used, now is predicted. The distance acid will penetrate along the fracture is then estimated based on fracture geometry and other formation and fracturing parameters. A proposed procedure for these calculations is outlined below. As previously indicated, models used here were selected because they give reasonable results and can be used with only a hand calculator or slide rule. Other possible models are discussed in Chapter 6.

1. Predict the temperature at the perforations, and at the middle and end of the fluid pad, using techniques discussed in Chapter 6, Section 6.4. Use the temperature in the middle of the pad as the pad fluid temperature entering the fracture to predict fracture geometry. The fluid temperature at the perforations after injection of the pad can be used as the acid temperature entering the fracture. These calculations can be made only with computer programs, such as those referenced in Chapter 6; therefore, in the following example, injection temperatures will be given — not calculated.

2. Predict the dynamic fracture geometry created by injection of the pad fluid, using one of the techniques described in Chapter 6, Section 6.2. The hand calculation procedure of Geertsma and de Klerk will be used to illustrate the example.

3. Predict the distance that acid can penetrate along the fracture created by the pad fluid, using the technique developed by Nierode and Williams and described in Chapter 6, Section 6.5. Predictions should be made for different pad volumes and acid injection rates to determine the design that will maximize profit from the treatment.

Example Calculation

To illustrate the procedure for predicting the acid penetration distance, an example design calculation is shown for a well completed in a limestone formation at 7,500 ft. This formation has a conductivity, kh_n, of 25 md-ft with a net interval thickness, h_n, of 50 ft. The well produces an oil that at reservoir temperature (200°F) has a viscosity of 0.5 cp. Other reservoir properties are summarized in Table 7.2.

A treatment will be designed using Fluid A (a fictitious viscous fluid) as a pad fluid. We will assume that this fluid is viscous at reservoir conditions (about 60 cp at shear rates obtained during flow along the fracture) and can be pumped through well tubular goods at 10 bbl/min without exceeding the allowable surface pressure of 5,000 psi. Fluid A contains a fluid-loss additive at a concentration of 20 lb/1,000 gal, to reduce the rate of fluid loss to the formation. A 15-percent hydrochloric acid (HCl) containing 50 lb polyacrylamide polymer per 1,000 gal acid will be considered for the treatment. We also will assume that this acid can be pumped at a maximum rate of 10 bbl/min. Laboratory tests show that for the expected acid temperature and shear rate in the fracture, the average viscosity will be about 1.2 cp. Tests also show the viscosity of totally reacted acid as it flows into the formation is 1.7 cp. Other fluid properties are summarized in Table 7.3.

Prediction of Bottom-hole Injection Temperature — A detailed prediction of the fluid-injection temperature profile will not be presented in this example. Instead, we will assume a calculation similar to those discussed in Chapter 5, Section 5.2, which shows that after from 5,000 to 10,000 gal of fluid injection, the temperature at the perforations stabilizes at 150°F. The reader should consult Chapter 6, Section 6.4 for a detailed review of techniques to predict bottom-hole temperature during fluid injection.

Predict the Dynamic Fracture Geometry

The method for predicting the dynamic fracture geometry created by the pad fluid used in this example was presented in Chapter 6, Section 6.2. The results of calculations made there for the pad fluid, and results of similar calculations for acid without a fluid-loss additive, are summarized in Table 7.4. The volume of fluid injected, average fracture width, length of one wing of the fracture, and total fracture volume are included in this summary for completeness. Only the average fracture width is required for subsequent calculations.

Predict the Acid Reaction Distance

Results from each step in calculating the acid penetration

TABLE 7.2—FORMATION PROPERTIES FOR EXAMPLE LIMESTONE FORMATION

Depth	7,500 ft
Formation thickness	
Gross	50 ft
Net	50 ft
Fracture gradient	0.7 psi/ft
Permeability	0.5 md
Porosity	0.10
Young's modulus	6.45×10^6 psi
Poisson's ratio	0.25
Reservoir fluid properties	
Viscosity	0.5 cp
Density	52 lb/cu ft
Compressibility	0.0001 psi^{-1}
Reservoir temperature	200°F
Reservoir pressure	2,500 psi
Well spacing	40 acres
Well radius	0.5 ft

TABLE 7.3—TREATMENT CHARACTERISTICS FOR THE EXAMPLE LIMESTONE FORMATION

Pad Fluid (Fluid A)
Injection rate (set by surface pressure and tubing size) .... 10 bbl/min
Temperature at which fluid enters the fracture .... 150°F
Average viscosity during flow along the fracture .... 60 cp
Fluid-loss additive concentration .... 20 lb/1,000 gal
Fluid-loss characteristics
   Spurt volume, V_{spt} .... 0.07 gal/sq ft
   Fluid-loss coefficient .... 0.002 ft/min$^{1/2}$

Acid
Injection rate (set by surface pressure limit and tubing size) .... 10 bbl/min
Average viscosity for flow along the fracture
  (175°F, partially reacted acid)
  15-percent HCl (containing 50 lb polyacrylamide/1,000 gal) .... 1.2 cp
Viscosity of the reacted acid (200°F) .... 1.7 cp
Acid density
  15-percent HCl (containing 50 lb polyacrylamide/1,000 gal) .... 71.1 lb/cu ft
Fluid-loss characteristics
   Spurt volume, V_{spt} .... 0.07 gal/sq ft
   Fluid-loss coefficient, C (no fluid-loss additive) .... 0.007 ft/min$^{1/2}$
   C (with fluid-loss additive) .... 0.002 ft/min$^{1/2}$

distance (defined as the point along the fracture where $c/c_o = 0.1$) are given in Table 7.5 for 15-percent HCl injected at 10 bbl/min and for different pad volumes. This calculation involves the use of a model developed by Nierode and Williams and discussed in Chapter 6, Section 6.5.

Step 1. Calculate the average fluid-loss velocity along the fracture at each time of interest, using Eq. 7.10.[16]

$$\bar{v}_N = \frac{\pi C}{2\sqrt{t}}, \text{ ft/min.} \quad \dots \dots \dots \dots \dots \dots \quad (7.10)$$

This procedure is performed assuming the fluid-loss coefficient for the acid is equal to that for the pad fluid, given as 0.002 ft/min$^{1/2}$ (to estimate the maximum expected acid penetration distance), and for the fluid-loss coefficient for

the reacted acid, estimated to be 0.007 ft/min$^{1/2}$ (to estimate the acid penetration distance if an effective fluid-loss additive is not used in the acid). The fluid-loss coefficients used here are given in Table 7.3 and must be considered as examples only. Actual coefficients should be derived from fluid-loss tests discussed in Chapter 5, Section 5.3 and in Chapter 6, Section 6.3. Results of calculations for our example are given in Table 7.5 under the heading Average Fluid-Loss Velocity.

Step 2. Calculate the Reynolds number for acid flow into the fracture, using Eq. 7.11.

$$N_{Re} = \frac{2\rho \bar{v}_A{}^{\circ} w_w}{\mu} \quad \dots \dots \dots \dots \dots \dots \dots \quad (7.11)$$

TABLE 7.4—FRACTURE GEOMETRY (EXAMPLE LIMESTONE FORMATION)

Time (minutes)	Fluid Volume Injected (bbl)	Average Fracture Width (in.)		Fracture Length* (ft)		Fracture Volume** (cu ft)	
		Pad Fluid	Acid Alone	Pad Fluid	Acid Alone	Pad Fluid	Acid Alone
15	150	0.13	0.02	235	94	255	16
30	300	0.15	0.02	357	133	446	22
45	450	0.17	0.03	453	163	642	41
60	600	0.18	0.04	536	189	804	63

*Length of one wing of a rectangular vertical fracture.
**Volume of both wings of the fracture.

TABLE 7.5—SUMMARY OF ACID PENETRATION DISTANCE CALCULATION ($c/c_o = 0.1$) (Example Limestone Formation)

Time (minutes)	Average Fluid-Loss Velocity (ft/min)		Peclet Number, N_{Pe}		Dimensionless Acid Penetration Distance, L_{aD}	
	Minimum	Maximum	Minimum Fluid Loss	Maximum Fluid Loss	Minimum Fluid Loss	Maximum Fluid Loss
15	0.00081	0.0028	0.24	0.13	0.27	0.13
30	0.00057	0.0020	0.20	0.092	0.21	0.10
45	0.00047	0.0016	0.18	0.11	0.19	0.12
60	0.00041	0.0014	0.17	0.13	0.16	0.13

Time (minutes)	Leakoff Reynolds Number, N_{Re}*		Acid Penetration Distance, xL (ft)	
	Minimum Fluid Loss	Maximum Fluid Loss	Minimum Fluid Loss	Maximum Fluid Loss
15	0.018	0.0098	135	18
30	0.015	0.0070	145	20
45	0.014	0.0084	160	30
60	0.012	0.0098	192	37

Noting that the flow velocity into one wing of the fracture is

$$v_A{}^\circ = \frac{i}{2h_g w_w} \, ,$$

we find that the Reynolds number can be written as

$$N_{Re} = \frac{\rho i}{\mu h_g} \, . \quad\dots\dots\dots\dots\dots\dots\dots\dots \text{(7.12)}$$

For this example, the Reynolds numbers is

$$N_{Re} = \frac{\left(71.1 \, \frac{lb}{cu \, ft}\right) \left(10 \, \frac{bbl}{min} \times 5.614 \, \frac{cu \, ft}{bbl}\right)}{\left(1.2 \, cp \times 0.04 \, \frac{lb/ft\text{-}min}{cp}\right)(50 \, ft)}$$

$$= 1{,}663.$$

Step 3. Read the effective mixing coefficient for the Reynolds number determined in Step 2 from Fig. 6.8. For this example,

$$D_e{}^{(\infty)} = 2.8 \times 10^{-4} \, \frac{sq \, cm}{sec} \, ,$$

or,

$$D_e{}^{(\infty)} = 2.8 \times 10^{-4} \, \frac{sq \, cm}{sec} \times \frac{sq \, ft}{(30.48 \, cm)^2} \times \frac{60 \, sec}{min}$$

$$= 1.8 \times 10^{-5} \, \frac{sq \, ft}{min} \, .$$

If the formation were a dolomite, the mixing coefficient would be reduced by multiplying by a factor determined

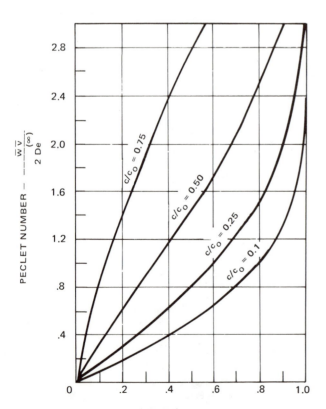

$$L_{aD} = \frac{2 \times L}{\overline{w}} \left(\frac{N_{Re}}{N_{Re}*}\right)$$

Fig. 7.14—Acid penetration distance along a fracture.[16]

from Fig. 6.9 to account for a reduced reaction rate at the fracture surface, as explained in Chapter 6.

Step 4. Calculate the Peclet number for fluid loss, using Eq. 7.13.

$$N_{Pe*} = \frac{\overline{w} \, v_N}{2 D_e{}^{(\infty)}} \, . \quad\dots\dots\dots\dots\dots\dots \text{(7.13)}$$

Results of this calculation are shown in Table 7.5. For the first entry, the value is calculated as follows:

$$N_{Pe*} = \frac{\left(0.13 \, in./12 \, \frac{in.}{ft}\right)(0.00081 \, ft/min)}{2(1.8 \times 10^{-5} \, sq \, ft/min)}$$

$$= 0.24.$$

Step 5. Read values for the dimensionless acid penetration distance from Fig. 7.14 (development of this figure is discussed in Section 6.5). Values are tabulated in Table 7.5.

Step 6. Calculate the acid penetration distance, xL, from the dimensionless distance using Eq. 7.14.

$$L_{aD} = \frac{2xL}{\overline{w}} \left(\frac{N_{Re}*}{N_{Re}}\right) \, , \quad\dots\dots\dots\dots \text{(7.14)}$$

or, solving for xL,

$$xL = \frac{\overline{w} \, L_{aD}}{2} \left(\frac{N_{Re}}{N_{Re}*}\right) \, . \quad\dots\dots\dots\dots \text{(7.15)}$$

Before making this calculation, it is necessary to calculate values for the fluid-loss Reynolds number, $N_{Re}*$. The definition of $N_{Re}*$ is

$$N_{Re*} = \frac{2\overline{w} \, v_N \, \rho}{\mu} \, .$$

Calculated values are given in Table 7.5. This calculation is illustrated next for the first entry in the table.

$$N_{Re*} =$$

$$\frac{2(0.13 \, in./12 \, in./ft)(0.00081 \, ft/min)(71.1 \, lb/cu \, ft)}{1.7 \, cp \left(0.04 \, \frac{lb/ft\text{-}min}{cp}\right)}$$

$$= 0.018.$$

The acid penetration distance can now be calculated using Eq. 7.15. Calculated values given in Table 7.5 were obtained as illustrated next for the first table entry.

$$xL = \frac{0.13 \, in. \, (0.27)(1{,}663)}{2(12 \, in./ft)(0.018)} \, ,$$

$$xL = 135 \, ft.$$

Results from the acid penetration calculations are summarized in Table 7.6 and compared with the fracture length. Included in this table are the maximum and minimum penetration distances for 15-percent HCl injected at 10 bbl/min. Penetration distances are listed for pad fluid volumes of 150, 300, 450, and 600 bbl.

7.5 Prediction of Fracture Conductivity and Stimulation Ratio

After predicting the acid penetration distance for the fluids of interest, complete the design as follows.

1. Predict the expected fracture conductivity from the

volume of acid to be used. If the conductivity is low, specify an acid volume sufficient to create an adequate conductivity.

2. Predict the stimulation ratio for each combination of pad and acid volume considered.

3. Select the most economic treatment (that is, the treatment that will best satisfy current guidelines for workover and stimulation projects, see Chapter 12 for guidelines).

Predict the Fracture Conductivity

The model described by Nierode and Kruk (see Chapter 6, Section 6.6) will be used here to predict the fracture conductivity resulting from acid reaction. To use this model, the fracture length contacted with acid (xL), the volume of acid injected (it), the acid dissolving power (X), and the rock embedment strength (S_{RE}) must be known. The theoretical fracture conductivity (assuming uniform reaction and no fracture closure) then can be predicted with Eq. 7.16, and the expected conductivity with Eq. 7.17.

$$wk_{fi} = 9.36 \times 10^{13} \left(\frac{w_a}{12}\right)^3, \quad \dots \dots \dots (7.16)$$

where $w_a = Xit/2xLh_g(1-\phi)$ with parameters defined to give w_a in inches and wk_{fi} in millidarcy-inches.

$$wk_f = C_1 \exp(-C_2\,\sigma), \quad \dots \dots \dots (7.17)$$

where $C_1 = 0.265\,(wk_{fi})^{0.822}$,

$C_2 = \left[19.9 - 1.3 \ln S_{RE}\right] \times 10^{-3}$

for $0 < S_{RE} < 20,000$ psi

and

$C_2 = \left[3.8 - 0.28 \ln S_{RE}\right] \times 10^{-3}$

for $20,000 < S_{RE} < 500,000$ psi

σ = closure stress = (fracture gradient × depth) − producing bottom-hole pressure.

One important step in this calculation is the selection of the acid volume for consideration. Based on experience, we recommend at least an acid volume 1.5 times the fracture volume contained between the wellbore and the acid penetration distance (predict this volume with Eq. 7.18), if 28-percent HCl is used.

$$V_f = it = 2xLh_g\bar{w}. \quad \dots \dots \dots (7.18)$$

If 15-percent HCl is used, an acid volume at least three times this fracture volume is recommended. When an effective fluid-loss additive is used in the acid, more acid than this minimum volume will be required to give an adequate con-

ductivity to maximize stimulation.

After setting the acid volume and predicting the fracture conductivity, determine the stimulation ratio from Fig. 7.15.

Example Calculation

To illustrate the calculation of fracture conductivity and stimulation ratio, the example design problem previously discussed will be continued. Values of parameters required for conductivity calculations are obtained as follows.

Acid penetration distance, xL. Tabulated in Table 7.6.

Acid dissolving power, $X_{15\ percent}$. From Table 3.4,

$$X_{15\ percent} = 0.082 = \frac{\text{cu ft rock dissolved}}{\text{cu ft acid injected}}.$$

Acid volume injected ($V = it$). Calculate the minimum recommended acid volume for each case of interest. In this example an acid volume three times the fracture volume contacted by acid following a pad fluid is used for all designs. This choice was made to indicate the effect of acid volume per unit of fracture length on fracture conductivity.

The minimum acid volume for the first entry in Table 7.6 is calculated as follows:

$$V_{15\ percent} = 3V_f = 3\left[2(135\ \text{ft})\,\frac{(50\ \text{ft})(0.13\ \text{in.})}{12\ \text{in./ft}}\right]$$

$$= 438\ \text{cu ft or }78.2\ \text{bbl}$$

Rock embedment strength. The rock embedment strength would normally be determined from measurements on formation cores. Rock strength was not given for the example, so a typical value of 50,000 psi will be assumed. Data given in Table 6.5 show that this embedment strength is typical of many limestones (that is, Desert Creek, Cisco, Edwards, Indiana, Pennsylvania, and Wolfcamp).

Closure stress (σ). The closure stress is equal to the fracture gradient times the formation depth minus producing bottom-hole pressure. For this example, we will assume the producing bottom-hole pressure is 1,500 psi; therefore, the closure stress is

TABLE 7.6—SUMMARY OF CALCULATED ACID PENETRATION
DISTANCE ($c/c_o = 0.1$)
(Example Limestone Formation)

Injected Volume (bbl)	Fracture Length (ft) Pad Fluid	Fracture Length (ft) Acid Alone	Acid Penetration Distance (ft) Minimum Fluid Loss	Acid Penetration Distance (ft) Maximum Fluid Loss
150	235	94	135	18
300	357	133	145	20
450	453	163	160	30
600	536	189	192	37

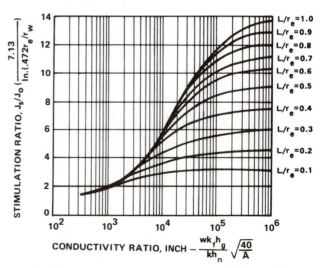

Fig. 7.15—Stimulation ratio from a vertical fracture.[21]

$\sigma = (0.7 \text{ psi/ft} \times 7{,}500 \text{ ft}) - 1{,}500 \text{ psi},$

$\sigma = 3{,}750 \text{ psi}.$

Step 1. Calculate the fracture width, w_a, as given by Eq. 6.36:

$$w_a = \frac{Xit}{2xLh_g(1-\phi)}.$$

For the example of interest,

$$w_a = \frac{0.082 \dfrac{\text{cu ft rock}}{\text{cu ft acid}} (it)\text{cu ft}}{2(50 \text{ ft})(1-0.1)xL \text{ ft}} = \frac{0.00091 \, (it)}{xL} \text{ ft}.$$

Values for it can be obtained from Column 2, Table 7.7 and those for xL from Column 5, Table 7.5. For the first entry in Column 3, Table 7.7, the calculation for w_a gives

$$w_a = \frac{0.00091 \, (438)}{135} = 0.0030 \text{ ft}.$$

Step 2. The ideal conductivity for Eq. 7.16 for the first entry is, therefore,

$wk_{fi} = 9.36 \times 10^{13} \, (0.0030)^3$

$\qquad = 9.36 \times 10^{13} \, (27 \times 10^{-9})$

$\qquad = 2.50 \times 10^6 \text{ md-in.}$

Step 3. Calculate the coefficients C_1 and C_2 for use in Eq. 7.17 to predict the expected fracture conductivity. The next example is for the first entry in Table 7.7. Other calculated values are in Table 7.7.

$C_1 = 0.265 \, (2.5 \times 10^6)^{0.822},$

$\qquad = 4.8 \times 10^4.$

Since $S_{RE} > 20{,}000$ psi,

$C_2 = \left[3.8 - 0.28 \ln \ (50{,}000)\right] \times 10^{-3},$

$\qquad = 0.77 \times 10^{-3}.$

TABLE 7.9—SCALING PARAMETERS FOR FIG. 7.15

Well Spacing (acres)	Drainage Radius, r_e (ft)	$\sqrt{\dfrac{40}{A}}$	$\dfrac{7.13}{\ln(0.472\,r_e/r_w)}$
10	330	2.0	1.11
20	467	1.42	1.05
40	660	1.0	1.00
80	933	0.71	0.95
160	1,320	0.50	0.91
320	1,867	0.35	0.87
640	2,640	0.25	0.84

Step 4. The expected fracture conductivity is obtained for calculated values of C_1 and C_2 as illustrated next for the first entry in Table 7.7.

$wk_f = 4.8 \times 10^4 \exp \ \left[-(0.77 \times 10^{-3})(3{,}750 \text{ psi})\right],$

$wk_f = 4.8 \times 10^4 \, (0.0557) = 2{,}700 \text{ md-in.}$

Other calculated values are given in Table 7.7.

Step 5. Calculate the fracture conductivity ratio, $(wk_f hg/kh_n)\sqrt{40/A}$. For the first entry in Table 7.7,

$$\frac{wk_f h_g}{kh_n} \sqrt{40/A} = \frac{2{,}700 \text{ md-in.}}{0.5 \text{ md}} \cdot \frac{50 \text{ ft}}{50 \text{ ft}} \sqrt{\frac{40 \text{ acres}}{40 \text{ acres}}}$$

$$= 5{,}400 \text{ in.}$$

Step 6. Calculate the fracture-length-to-drainage-radius ratio, L/r_e. For 40-acre spacing, $r_e = 660$ ft (see Table 7.9). For Entry 1 in Table 7.8,

$$\frac{xL}{r_e} = \frac{135 \text{ ft}}{660 \text{ ft}} = 0.20.$$

Step 7. Read the stimulation ratio from Fig. 7.15. For the first entry,

$J_s/J_o = 3.2$, for 40-acre spacing.

If spacing were not 40 acres, the scaling factor, $\left[7.13/\ln(0.472 \, r_e/r_w)\right]$, must be calculated and divided into the value read for the ordinate of Fig. 7.15. Values for this

TABLE 7.7—SUMMARY OF FRACTURE CONDUCTIVITY CALCULATION
(Example Limestone Formation)

Pad Volume (bbl)	Minimum Acid Volume (cu ft)	Fracture Width, w_a (ft)		Ideal Fracture Conductivity, wk_{fi} (md-in.)	
		Minimum Fluid Loss	Maximum Fluid Loss	Minimum Fluid Loss	Maximum Fluid Loss
150	438 (78.2 bbl)	0.0030	0.022	2.5×10^6	1.0×10^9
300	544 (97.0 bbl)	0.0034	0.025	3.7×10^6	1.5×10^9
450	680 (121.2 bbl)	0.0039	0.021	5.6×10^6	0.87×10^9
600	864 (154.0 bbl)	0.0041	0.021	6.5×10^6	0.87×10^9

Pad Volume (bbl)	C_1		Fracture Conductivity, wk_f (md-in.)	
	Minimum Fluid Loss	Maximum Fluid Loss	Minimum Fluid Loss	Maximum Fluid Loss
150	4.8×10^4	6.6×10^6	2,700	370,000
300	6.6×10^4	9.2×10^6	3,700	510,000
450	9.3×10^4	5.9×10^6	5,200	330,000
600	10.6×10^4	5.9×10^6	5,900	330,000

TABLE 7.8—SUMMARY OF STIMULATION RATIO CALCULATION

Pad Volume (bbl)	Acid Volume (bbl)	$\dfrac{wk_f h_g}{k h_n} \sqrt{\dfrac{40}{A}}$		xL/r_e		J_s/J_o	
		Minimum Fluid Loss	Maximum Fluid Loss	Minimum Fluid Loss	Maximum Fluid Loss	Minimum Fluid Loss	Maximum Fluid Loss
150	78	5,400	740,000	0.20	0.03	3.2	2.0
300	97	7,400	1,020,000	0.22	0.03	3.6	2.1
450	121	10,400	660,000	0.24	0.05	4.0	2.3
600	154	11,800	660,000	0.29	0.06	4.4	2.5

group are tabulated in Table 7.9 for typical well spacings.

For situations where the fracture length is less than one-tenth of the drainage radius, such as the example cases for maximum fluid loss, Fig. 7.15 cannot be used. For these cases, Eq. 7.20, derived by Raymond,[22] can be used to obtain an approximate value for J_s/J_o.

$$J_s/J_o = \frac{\ln (r_e/r_w)}{\ln \left\{ \dfrac{xL\pi + \dfrac{wk_f h_g}{k h_n}}{\dfrac{wk_f h_g}{k h_n}} \right\} + \ln \dfrac{r_e}{xL}} \qquad \ldots (7.20)$$

For the first entry in the maximum fluid-loss case, the stimulation ratio is calculated as follows:

$$\frac{J_s}{J_o} = \frac{\ln (660 \text{ ft}/0.5 \text{ ft})}{\ln \left(\dfrac{(12)(18)\pi + 370{,}000}{370{,}000} \right) + \ln \left(\dfrac{660 \text{ ft}}{18 \text{ ft}} \right)}$$

$$= \frac{7.18}{0.0018 + 3.6} = 2.0.$$

Note that Eq. 7.20 is approximate and predicts somewhat higher stimulation ratios for low fracture conductivity ratios than obtained from Fig. 7.15.

An investigation of results tabulated in Table 7.8 can help develop an understanding of factors limiting stimulation from acid fracturing treatments. Consider the following:

1. The stimulation ratio was larger when a pad was used and followed by acid containing an effective fluid-loss additive than when acid alone was injected. For example, $J_s/J_o = 3.2$ when a 150-bbl pad was used, followed by 78 bbl of acid with an effective fluid-loss additive; 2.0 when 78 bbl of acid without an additive was injected without a pad.

2. An improved treatment could be obtained with the pad fluid when an effective fluid-loss additive was used in the acid if more than the minimum volume (78 bbl) of acid was injected. For example, if after the 150-bbl pad, enough acid was injected to give a conductivity ratio of 740,000 (the same as for the short fracture), a stimulation ratio of 4.5 was expected (read from Fig. 7.15 for $L/r_e = 0.2$, $wk_f/k = 740{,}000$). This would require an acid volume of 584 bbl or 24,528 gal. The required acid volume was calculated as follows:

$$78 \text{ bbl} \times \frac{135 \text{ ft}}{18 \text{ ft}} = 584 \text{ bbl} (24{,}528 \text{ gal})$$

This calculation shows that it is possible to optimize the treatment design to maximize stimulation at minimum cost by considering different combinations of acid and pad volumes. Also, the concept that the acid volume required to give maximum stimulation will increase as the acid penetration distance increases is apparent.

3. The stimulation ratio will vary with well spacing. The treatment with a 150-bbl pad and 78 bbl of acid, for example, would give the following scaling groups, if the well to be treated were completed on 640-acre spacing.

$$\frac{wk_f h_g}{k h_n} \sqrt{\frac{40}{A}} = 5{,}400 \, (0.25) = 1{,}350 \text{ in.,}$$

$$\frac{L}{r_e} = \frac{135 \text{ ft}}{2{,}640 \text{ ft}} = 0.05.$$

From Fig. 7.15, we estimate that $J_s/J_o \left[7.13/\ln(0.472 \, r_e/r_w) \right] = 2.0$ for these values of fracture-conductivity ratio and dimensionless fracture length. Noting that the scaling factor has a value of 0.84 for 640-acre spacing (from Table 7.9), the expected stimulation ratio is 2.38 as

$$J_s/J_o = \frac{2.0}{0.84} = 2.38.$$

4. The stimulation ratio indicated in Table 7.8 is expected only if the well is undamaged and is not stimulated before fracturing. If the well is damaged before the fracturing treatment, the observed production ratio will be equal to that expected for damage removal (see Chapter 2, Section 2.3) multiplied by J_s/J_o for the fracturing treatment. If the well was stimulated previously, and is still producing at a stimulated rate, the observed J_s/J_o will be less than calculated. To estimate the stimulation from a fracturing treatment in a stimulated well, divide the theoretical J_s/J_o by the extent to which the well was stimulated at the time the second treatment was conducted.

5. The stimulation ratio calculated here is the stabilized, steady-state increase in productivity. Because of the high rate often seen immediately after stimulation, production data for use in evaluating treatment economics should be taken after production has stabilized following the treatment. In low permeability formations, flow may not stabilize for several days or weeks following the treatment.

7.6 Select the Most Economic Treatment

The previously described design procedure should be carried out for all treatments that appear to have promise in a given reservoir. Ideally, the cost of each treatment should be computed and the treatment offering the maximum stimulation at minimum cost selected. An economic analysis of this or other treatments should follow general practices for evaluation of workover or stimulation candidates. Techniques for economic analysis are discussed in Chapter 12.

7.7 Design Hints for Unconventional Treatments

As indicated, one main requirement for a successful acid fracturing treatment is the control of acid fluid-loss rate to the formation. Unless the rate of acid fluid loss is controlled, viscous prepads, precooling pads, or chemically retarded acids are of minimal benefit. When designing a treatment to use any of these special procedures or additives, consider the following:

Acid fluid-loss additives. Not all products marketed as acid fluid-loss additives are effective. Before using an additive, therefore, data demonstrating the effectiveness of the additive should be required. In these tests, acid-containing additive should be injected into a 12- to 18-in. long *carbonate* core similar to the formation to be fractured, the differential pressure across the core should be approximately equal to the filtration-pressure difference expected in the treatment (fracture gradient × depth − reservoir pressure), and *both* the acid and the core should be at reservoir temperature. The C_w values for the additive should then be obtained from the test results. In the past, erroneous laboratory data have been obtained by using sandstone cores, by injecting cool acid into a core preheated to reservoir temperature, and by conducting tests at low differential pressures.

Once accurate test data are available, results from the treatment should be predicted using a fluid-loss coefficient for the acid containing the fluid-loss additive (thereby neglecting the effect of the pad fluid). This will be the minimum possible acid penetration distance. The penetration distance calculated, assuming the fracture geometry does not differ from that created by the pad, is often optimistic, because acids seldom have as low a fluid-loss rate as the pad fluid.

The main difficulty anticipated with this procedure is scaling up fluid-loss data from a test conducted on a small-diameter core to a fracture with a surface area of several hundred square feet. Problems should be expected because we are assuming (1) that the wormhole spacing in the field will be the same as in the small core in the lab and (2) that the fracture does not intersect fractures or vugs in the reservoir too large to be bridged by the fluid-loss additive. In reality, the number of wormholes created in the field is unknown and carbonate formations are often vugular or naturally fractured.

This procedure for evaluating acid fluid-loss additives is not effective for naturally fractured formations. Fluid-loss additives may be of some benefit in this type formation; however, there is no known way to quantify their effect. In some instances, the effectiveness of the additive in a fractured formation possibly can be enhanced by adding a finely ground silica sand (about 70- to 300-mesh range) at a concentration of 20 to 40 lb/1,000 gal of acid in an attempt to plug natural fractures intersecting the hydraulically induced fracture.

Emulsified Acids. When designing treatments using an emulsified acid (either oil or acid external), the emulsion viscosity in the fracture should be used to estimate the acid penetration distance. This viscosity should be evaluated at the expected temperature and shear rate in the fracture after partial reaction of the acid. The rate of fluid loss should be estimated using data from the fluid-loss test procedure described previously. When conducting the fluid-loss test, the temperature and reaction history the emulsified acid will see in the field must be simulated. If this is not done, emulsion properties often will not correspond to the properties observed in the treatment, and predictions can be substantially in error.

Viscous and Controlled Density Pad Fluids. As discussed in Chapter 5, viscous pad fluids are sometimes selected to promote acid channeling through the pad fluid to maximize penetration distance and fracture conductivity. More or less dense pad fluids are sometimes selected to promote acid flow along the top or bottom of a fracture.[23] None of these techniques have proved successful in routine field application now, although they appear promising.

When designing a treatment with these novel techniques, we recommend the assumption that acid contacts one-third to one-half of the vertical fracture when calculating the acid penetration distance. Assuming that acid fluid losses occur over the total permeable thickness is preferred for fluid-loss calculations.

Precooling pad. A large pad of water has often been used to cool the formation before acid injection in an effort to increase the acid penetration distance. Normally, in limestone and dolomite formations at temperatures above about 200°F, a precooling pad is of only marginal benefit (see discussion in Chapter 5), since the acid penetration along the fracture is often limited by the rate of acid fluid loss or by the rate of acid transfer to the fracture walls, not the reaction rate at the fracture walls. When properly designed, a precooling pad may be effective for cooling well-tubular goods, thereby allowing concentrated acids to be used in deep hot formations.

Chemically retarded acids have been developed in an attempt to reduce acid reaction rate and increase the acid penetration distance. Dynamic reaction rate tests described in Chapter 5 show that these retarded acids have only marginal benefit.

7.8   Design of Acid Fracturing Treatments To Remove Near-Wellbore Damage

An acid fracturing treatment to remove near-wellbore damage can be designed to treat the total interval using one of many techniques, depending on the type of completion. In general, however, all designs can be classified either as a staged or limited entry technique. In these treatments, a relatively small volume of acid (50 to 200 gal/ft of interval) is injected into the formation at a high rate. The fracture created will be only a few feet long, and the stimulation ratio will be that attained from damage removal alone.

Staged Treatments

In a staged treatment, the well is treated with a volume of acid followed by material to divert fluid flow from the interval just fractured to another section of the wellbore. Diversion can be attained in open-hole completions using large-grain, oil-soluble solids or rock salt. In a cased well, diversion often is obtained more easily if the more productive sections are completed with an equal number of perforations in each interval. The acid treatment can then be run in equal volume stages with stage separation by ball sealers or diverting material. Enough ball sealers or diverting agent should be injected following each stage to close the perforations in one interval. In general, an acid volume of about 50 to 200 gal/ft of gross interval will be adequate to remove damage. A detailed discussion of diverting agents is given in Chapter 11, Section 11.7.

If, for example, a well is to be completed in a carbonate section that has a 500-ft gross interval and porosity logs show five zones with an average thickness of 25 ft, the following completion-stimulation technique is favored. (1) Case the well and perforate with an equal number of perforations in each interval (often about 10 per interval). (2) Design the stimulation treatment for the average zone thickness and pump five identical stages separated by ball sealers (or diverting agent, if preferred). If it is not economical to perform a large treatment to give stimulation in addition to damage removal, damage often can be removed with a design similar to that proposed below.

Step 1. Inject 2,500 gal (100 gal/ft × 25 ft) of 15-percent HCl at the maximum allowable rate without exceeding the surface pressure limit. During the last 1,500 gal, drop one ball sealer every 150 gal (total of 10 balls).

Steps 2 through 4. Repeat Step 1.

Step 5. Repeat Step 1 — start ball sealers after 1,000 gal and drop 1 every 300 gal (total of 5 balls).

Step 6. Overdisplace with 500 gal of brine.

Step 7. Place the well on production.

The cased completion is preferred to an open-hole completion because it allows better control of gas or water production from the different intervals, if or when it occurs, and stimulation of the interval is usually simplified.

The Limited Entry Technique

The limited entry technique, or a modified limited entry technique, is often used to remove damage in massive carbonate formations.[24, 25] In this treatment, the well is completed with an unusually low-perforation density, often with only one 0.25-in.-diameter perforation in each 20- to 100-ft interval. The well is perforated with a limited number of small diameter perforations so that the friction pressure for flow through the perforations will hold the pressure in the casing above the breakdown pressure for other intervals. In theory, the well can be treated with a single stage. In practice, a modified treatment is often used in which the well is treated in stages by injecting a given volume of acid, dropping a ball sealer, injecting another volume of acid, and repeating the process until all perforations have been temporarily sealed and the total interval has accepted acid.

Design of a modified limited entry acid fracturing treatment is still somewhat of an art, since the number of perforations accepting fluid at any time is seldom known. However, a treatment consisting of about 30 to 100 gal of acid per foot of interval is normal. This treatment should be staged so that all, or almost all, the perforations are sealed with ball sealers by the end of the procedure. A design procedure for a limited entry treatment was given by Howard and Fast[26] in a previous monograph and will not be repeated here.

References

1. Sarmiento, R.: "Geological Factors Influencing Porosity Estimates from Velocity Logs," *Bull.*, AAPG (May 1961) **45**, No. 5, 633-644.

2. Beal, Carlton: "The Viscosity of Air, Water, Natural Gas, Crude Oil and its Associated Gases at Oil Field Temperatures and Pressures," *Trans.*, AIME (1946) **165**, 103-115.

3. Chew, Ju-Nan and Connally, C. A., Jr.: "A Viscosity Correlation for Gas Saturated Crude Oils," *Trans.*, AIME (1959) **216**, 23-25.

4. Dorsey, N. E.: "Properties of Ordinary Water Substances in all 'its' Phases," *Monograph Series No. 81*, ACS (1940).

5. *International Critical Tables*, McGraw-Hill Book Co., Inc., New York (1926).

6. *Engineering Data Book*, Natural Gas Processors Suppliers Assn., Tulsa (1977).

7. Truse, A. S.: "Compressibility of Natural Gases," *Trans.*, AIME (1957) **210**, 355-357.

8. Standing, M. B.: *Volumetric and Phase Behavior of Oil Field Hydrocarbon Systems*, Society of Petroleum Engineers of AIME, Dallas (1977).

9. *Frac Guide Data Book*, Dowell, Tulsa.

10. *The Fracbook Design/Data Manual*, Halliburton Services Ltd., Houston.

11. *Product Technical Guide*, The Western Co., Fort Worth.

12. *Applied Engineering Stimulation*, BJ Hughes, Houston.

13. Kiel, O. M. and Weaver, R. H.: "Emulsion Fracturing System," *Oil & Gas J.* (Feb. 1972) **70**, 72-73.

14. Sinclair, A. R., Terry, W. M., and Kiel, O. M.: "Polymer Emulsion Fracturing," *J. Pet. Tech.* (July 1974) **26**, 731-738.

15. Kucera, C. H.: "New Oil Gelling Systems Prevent Damage in Water Sensitive Sands," paper SPE 3503 presented at the SPE-AIME 46th Annual Meeting, held in New Orleans, Oct. 3-6, 1971.

16. Williams, B. B. and Nierode, D. E.: "Design of Acid Fracturing Treatments," *J. Pet. Tech.* (July 1972) **24**, 849-859; *Trans.*, AIME, **253**.

17. Nierode, D. E., Bombardieri, C. C., and Williams, B. B.: "Prediction of Stimulation from Acid Fracturing Treatments," *J. Cdn. Pet. Tech.* (Oct.-Dec. 1972) 31-41.

18. Geertsma, J. and de Klerk, F.: "A Rapid Method of Predicting Width and Extent of Hydraulically Induced Fractures," *J. Pet. Tech.* (Dec. 1969) 1571-1581; *Trans.*, AIME, **246**.

19. Williams, B. B.: "Fluid Loss from Hydraulically Induced Fractures," *J. Pet. Tech.* (July 1970) 882-888; *Trans.*, AIME, **249**.

20. Broaddus, G. C. and Knox, J. A.: "Influence of Acid Type and Quantity in Limestone Etching," paper 851-39-I presented at the API Mid-Continent Meeting, Wichita, Kans., March 31-April 2, 1965.

21. McGuire, W. J. and Sikora, V. J.: "The Effect of Vertical Fractures on Well Productivity," *Trans.*, AIME (1960) **219**, 401-403.

22. Raymond, L. R. and Binder, G. G., Jr.: "Productivity of Wells in Vertically Fractured, Damaged Formations," *J. Pet. Tech.* (Jan. 1967) 120-130; *Trans.*, AIME, **240**.

23. Fredrickson, S. E. and Broaddus, G. C.: "Selective Placement of Fluids in a Fracture by Controlling Density and Viscosity," *J. Pet. Tech.* (May 1976) 597-602.

24. Murphy, W. B. and Juch, A. A.: "Pin-Point Sand-Fracturing — A Method of Simultaneous Injection into Selected Sands," *J. Pet. Tech.* (Nov. 1960) 21-24.

25. Lagrone, K. W. and Rasmussen, J. W.: "A New Development in Completion Methods — The Limited Entry Technique," *J. Pet. Tech.* (July 1963) 695-702.

26. Howard, G. C. and Fast, C. R.: *Hydraulic Fracturing*, Monograph Series, Society of Petroleum Engineers of AIME, Dallas (1970) **2**.

Chapter 8

Models for Matrix Acidizing

8.1 Introduction

Several mathematical models have been developed to describe changes that occur when acid is injected into a porous medium. This chapter describes the models for matrix acidization. These models consider both the modification of the pore structure as it dissolves and the change in acid concentration as a function of both time and position within the pore system. Models applicable to both carbonate and sandstone acidizing are included.

8.2 Description of a Model for Matrix Acidization

The extent to which acid penetrates a rock matrix depends on both the matrix properties and the local reaction rate. The reaction rate in turn depends on matrix properties and other variables, such as the temperature, pressure, and composition of the reacting fluids. One of the essential matrix properties is that of the microstructure of the porous material invaded by the acid. Therefore, it is necessary to characterize the pore structure and to determine the change in this structure as acid reaction proceeds. The geometry of a real system is very complex and defies precise mathematical description. The only way relevant variables can be isolated and studied at this time is within the formal structure of a model. This approach has been used by many investigators in an attempt to attach physical meaning to parameters used to correlate the behavior of fluids in porous media. Dullien[1] presented a comprehensive literature survey of the models and methods used to determine pore-size distributions in a porous medium.

Scheidegger[2] reviewed capillary models and found that to predict those quantities related to the geometric structure of a porous medium, such as permeability and capillary pressure, an empirical correction factor (often called the tortuosity) must be introduced. This same conclusion has been verified by a number of other investigators.[3-6]

The model proposed by Schechter and Gidley[7] to describe matrix acidization is a capillary model. It is, however, used in a limited manner so that the quantities of interest are taken in the form of ratios. Therefore, it is not necessary to predict the absolute value of the properties. It is suggested that when used in this limited sense, the model will be subject to fewer uncertainties than other investigators encountered.

In this model, pores are assumed to be interconnected so that a fluid can flow through the matrix under the influence of a pressure gradient. The number of capillaries per unit volume having a cross-sectional area between A and $A + dA$ is represented by a function, $\eta(A,X,t)dA$. The notation indicates that the number of pores of a given size may differ at different positions within the matrix, X, and at a given position with time, t.

Given the distribution function, η, this model predicts that the total number of pores per unit volume at any point and any time is

$$n = \int_0^\infty \eta(A,X,t)\, da, \quad \dots\dots\dots\dots\dots\dots (8.1)$$

and the porosity is

$$\phi = \ell \int_0^\infty A\, \eta(A,X,t)dA. \quad \dots\dots\dots\dots\dots (8.2)$$

Here, ℓ is the average length of a pore.

As acid reacts within the matrix and the pores increase in size, the function η will change. This evolution in pore size depends on two factors: (1) the gross configuration of the matrix and (2) the rate at which a given pore is enlarged by the acid attack. The configuration of the matrix during acidization is described by the distribution function, $\eta(A,X,t)dA$. The over-all reaction rate depends largely on the nature of the surface reaction within a pore; however, in most cases, it also depends on the pore size, matrix, and other properties. Thus, these two factors are coupled.

The capillary model allows the permeability ratio (that is, the permeability of the porous medium at any time during acidization, k, divided by the original permeability, k_o) to be represented by Eq. 8.3.

$$k/k_o = \frac{\int_0^\infty A^2\, \eta(A,X,t)dA}{\int_0^\infty A^2\, \eta(A,X,0)dA}. \quad \dots\dots\dots\dots (8.3)$$

To simulate the acidizing process, we see from Eqs. 8.1 through 8.3 that the change in the pore size distribution, $\eta(A,X,t)$, as a function of time must be known. Two fundamental processes permit a change in pore size: (1) reaction with the walls of a pore and (2) interconnection of pores when the solid separating the flow channels is dissolved. To describe the rate of change in area of a single pore, a function, $\psi(A,X,t)$, called the pore growth function, is defined.

$$\psi(A,X,t) = \frac{dA}{dt} . \quad \dots\dots\dots\dots\dots \quad (8.4)$$

The function form for ψ, which is determined by the surface reaction kinetics and the pore geometry, will be defined later.

The two mechanisms for changes in the pore size distribution can be combined as follows.

$$\frac{\partial \eta}{\partial t} + \frac{\partial}{\partial A}(\psi\eta) = \ell \left\{ \int_0^A \eta(A-\lambda,X,t) \right.$$

$$\eta(\lambda,X,t)\,\psi(\lambda,X,t)d\lambda - \int_0^\infty \left[\psi(A,X,t) \right.$$

$$\left. + \psi(\lambda,X,t) \right] \eta(A,X,t)\,\eta(\lambda,X,t)d\lambda \left. \right\} . \quad \dots\dots \quad (8.5)$$

The derivation of this expression is straightforward, but lengthy. Details may be found elsewhere.[7,8]

Eq. 8.5 shows that the change in pore size distribution is given by two distinct types of terms: (1) terms linear in η, which represent the change brought about by a single pore increasing in diameter and (2) quadratic terms that represent interconnection of pores when the separating wall is dissolved.

The pore growth function, ψ, is very important in the acidizing process. This parameter can be related to reaction kinetics by writing a material balance on a single pore as

$$\rho_{ma} \ell \frac{dA}{dt} = r_{ave} \beta \Gamma \ell, \quad \dots\dots\dots\dots \quad (8.6)$$

or

$$\psi = \frac{r_{ave} \beta \Gamma}{\rho_{ma}} . \quad \dots\dots\dots\dots \quad (8.7)$$

Here r_{ave} is the average wall reaction rate taken over the entire reactive surface area (defined as the mass of acid reacted per unit area per unit time). The quantity β is the mass of rock dissolved per mass of acid reacted. Values of formation density, ρ_{ma}, are given in Table 3.4. Γ is the circumference of the pore (for a circular pore, $\Gamma = 2\sqrt{\pi A}$) and ℓ is the pore length. The term, $r_{ave} \beta \Gamma \ell / \rho_{ma}$, therefore, represents the volume of rock dissolved per unit time. Values for the average reaction rate, r_{ave}, used to predict ψ may be obtained using techniques described in Chapter 4. In this regard, it should be noted that R^*, the dimensionless reaction rate, is related to r_{ave} by the following equation:

$$R^* = \Gamma \ell\, r_{ave} / c_o q, \quad \dots\dots\dots\dots \quad (8.8)$$

where q is the volumetric flow rate in the pore.

In this discussion of modeling of matrix acidizing, it should be clear that the same surface kinetic model applies to both matrix acidization and acid fracturing. The hydrodynamics and geometry within which reaction occur are different, however. As in the fracturing treatment, the geometry within which reaction occurs must be known.

Eqs. 8.2 and 8.3 relate the porosity and permeability to integrals of the pore size distribution. To simplify this calculation it has been shown[7,8] that these relationships can all be expressed in the form of moments as defined by Eq. 8.9 (the jth moment):

$$M_j = \int_0^\infty A^j\, \eta(A,X,t)dA. \quad \dots\dots\dots\dots \quad (8.9)$$

Substituting into Eqs. 8.2 and 8.3 yields

$$\phi = \ell M_1, \quad \dots\dots\dots\dots\dots \quad (8.10)$$

and

$$\frac{k}{k_o} = \frac{M_2(X,t)}{M_2(X,0)} . \quad \dots\dots\dots\dots \quad (8.11)$$

For some growth functions, it is possible to solve directly for the moment using Eq. 8.5 or, if these special conditions are not satisfied, to obtain approximate solutions without ever solving for the pore size distribution. The reader should consult Refs. 7 through 9 for a more complete discussion of moments and how they can be used to simplify prediction of changes occurring during matrix acidization.

8.3 Application of the Matrix Acidizing Model to Retarded Acid Systems

To obtain a greater depth of penetration, various chemicals have been included in acid to retard the rate of reaction. It is of interest to determine the permeability distribution, and, hence, the increase in production, which is expected if the reaction rate is so highly retarded that essentially no reaction takes place during the time that acid is being pumped into the reservoir. This hypothetical system has been proposed as the ultimate for a matrix treatment.

On injecting a retarded acid, each pore will contain a mass of acid equal to the pore volume multiplied by the acid concentration, $\ell A c$, where c is the initial acid concentration. Given sufficient time, the acid in each pore will react and will dissolve, to some extent, the walls of that pore. Pores that have the largest area therefore will be enlarged to a greater extent than small pores because the acid present is proportional to the pore area.

Schechter and Gidley[7] have shown that for this case, one can solve directly for the moments with the result that

$$\frac{n}{n_o} = \frac{M_o}{M_o(0)} = \exp(\phi_o - \phi), \quad \dots\dots\dots \quad (8.12)$$

$$\frac{k}{k_o} = \left(\frac{\phi}{\phi_o}\right)^2 \exp\left[2(\phi - \phi_o)\right]. \quad \dots\dots\dots \quad (8.13)$$

Here, $M_o(0)$, n_o, and ϕ_o denote quantities evaluated initially, before acid is introduced.

To test this prediction, Guin $et\ al.$[9] performed experiments using cold ($\sim -30°C$) 48-percent hydrofluoric acid imbibed into precooled, fritted glass disks. At this temperature, the HF reaction rate with silica is very slow, closely modeling the use of a highly retarded acid. Permeabilities measured after acidization divided by the initial permeability of the fritted glass are plotted against the theoretical ratio in Fig. 8.1. Theoretical values are obtained from measured porosities of the acidized disks and the application of Eq. 8.13. The bulk of the data falls within ± 10 percent of the theoretical value (indicated by dashed lines). This comparison indicates the model simulates the behavior of the real system without using any adjustable parameters.

Note that the permeability increase predicted by Eq. 8.13 is small, even for relatively large porosity changes. Large stimulation ratios therefore will normally not be possible from matrix acidizing an undamaged formation with a retarded acid, even though the acid may penetrate a significant distance into the formation. The expected productivity change can be predicted with Eq. 2.1 if permeability after acidization and radius of acid reaction are known.

8.4 Application of the Model to Slowly Reacting Systems

A comprehensive application of the model to predict the changing porosity and permeability of a porous glass disk being dissolved by flowing hydrofluoric acid has been reported by Glover and Guin.[10] They solved the evolution equation (Eq. 8.5) approximately, using a polynomial expansion method that appears adequate if the reaction rate is slow. Experimentally determined porosity and permeability measured as a function of time were in excellent agreement with theoretical prediction. This comparison is significant since there were no adjustable parameters. The agreement between experiment and theory is shown in Figs. 8.2 and 8.3. The abscissa shown denotes a reduced time measured in microns that is proportional to time as indicated by Eq. 8.14.

$$\tau = \frac{2\sqrt{\pi}\,\beta\,\xi_f'\,c_o\,t}{\rho_{ma}}\,, \quad\ldots\ldots\ldots\ldots\ldots\ldots\ (8.14)$$

where β is the mass of rock per mass of acid as defined in Chapter 3, ξ_f' is the kinetic parameter (see Table 4.1), and ρ_{ma} is the density of the solid.

Sinex et al.[11] reported similar experiments in which a dish of porous copper was dissolved by a solution of ferric citrate. Again, the results were in agreement with model predictions (Fig. 8.4). The broad band represents the range of results obtained by solving Eq. 8.5 for a variety of initial pore size distributions.

8.5 Application of the Model to Fast-Reacting Systems

The solution of Eq. 8.5 is not easy when the pore growth

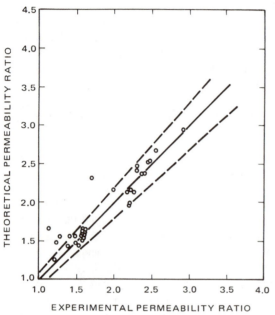

Fig. 8.1—Comparison of experimental with theoretical permeability ratios for a highly retarded acid.[9]

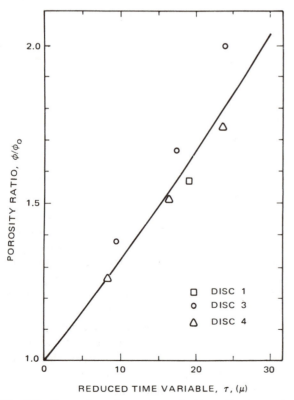

Fig. 8.2—Comparison of experimental with predicted porosity improvement in a porous glass disk during acidization.[10]

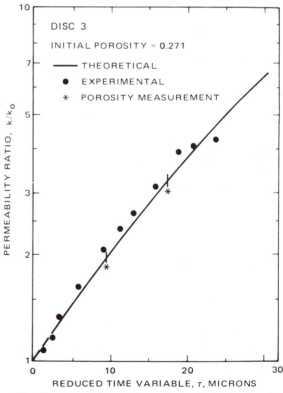

Fig. 8.3—Comparison of experimental with predicted permeability improvement in a porous glass disk during acidization.[10]

function, $\psi(A,X,t)$, is selected to represent fast reactions (that is, when diffusion is the rate-determining step for all but the smallest pores). Because of this difficulty, Guin and Schechter[12] used a numerical simulation technique based on Monte Carlo methods. Their study demonstrated that the model will predict qualitatively what occurs in carbonate acidizing, in that wormhole formation is predicted. The phenomenon of wormholing was found to be related to the surface reaction rate. Those systems for which the reaction rate is large will readily form wormholes, while slowly reacting systems do not.

Permeability increases predicted by Guin and Schechter are smaller than those observed in carbonate acidizing but are typical of those observed in more homogeneous sandstones. This result apparently occurred because only 1,000 pores could be included in the numerical calculation. Because of the small number of pores considered relative to the number present, one of the few large pores present in a naturally occurring carbonate seldom would be selected randomly from a representative population (that is, the pore that becomes the wormhole may be one out of 1 million).

Because a valid solution of Eq. 8.5 has not been obtained yet for the case of fast reactions and because the wormholing phenomenon is the most important feature of carbonate acidizing, a phenomenological model for wormhole growth, in which those parameters limiting the growth are identified, is the only approach for studying matrix acidizing of carbonates presently available. This treatment is presented in Section 8.7. Because sandstones react more slowly than carbonates, and, therefore, the tendency to wormhole is much less pronounced, the general model can be applied to sandstone acidizing.

8.6 Prediction of Acid Penetration Into Sandstones

Williams and Whiteley[13] and McCune et al.[14] have proposed methods for simulating the sandstone acidizing process. Both start with an acid balance but introduce different assumptions, thereby leading to differing results. These two models are discussed in this section.

By writing a mass balance on a pore and using the pore size distribution, one can show that the change in acid concentration as a function of distance and time can be related to a reaction-rate coefficient defined by the following equation:

$$\frac{\partial(\phi\bar{c})}{\partial t} + \bar{v}_A \frac{\partial\bar{c}}{\partial X} = -\phi\bar{c}\left[\widetilde{rf}(\phi)\right]. \qquad (8.15)$$

In this equation, $\bar{c}$ is the average concentration of acid at a position X measured from the inlet, u is the volumetric flux, and $\widetilde{rf}(\phi)$ is an effective reaction-rate coefficient that varies with the changes occurring during acidization. This form of the reaction rate was suggested to account for changes in pore size and porosity and for changes in reaction rate expected because clays react much more rapidly with HF (see Table 4.1, Chapter 4). An analytical solution to Eq. 8.15 can be obtained if one assumes that a steady state is developed and porosity is constant. For these assumptions, the mass balance equation reduces to

$$u \frac{\partial\bar{c}}{\partial X} = -\phi\bar{c}\left[\widetilde{rf}(\phi)\right]. \qquad (8.16)$$

Further, if the inlet condition of $c = c_o$ at $X = 0$ is inserted, Eq. 8.15 reduces to

$$\frac{\bar{c}(X)}{c_o} = \exp\left[\frac{-\phi\left[\widetilde{rf}(\phi)\right]X}{u}\right]. \qquad (8.17)$$

The reaction-rate coefficient, $\widetilde{rf}(\phi)$, can be determined from experimental measurements obtained by injecting hydrofluoric acid into a linear sandstone core and observing the effluent acid concentration. A typical effluent hydrofluoric-acid concentration profile is shown in Fig. 8.5. The effective reaction rate coefficient, $\widetilde{rf}(\phi)$, can be obtained from the data given in Fig. 8.5 using Eq. 8.17. Fig. 8.6 illustrates the variation in reaction-rate coefficient as a

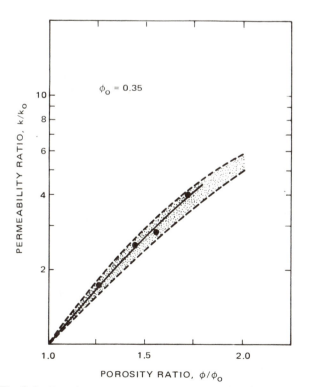

Fig. 8.4—Experimental results compared with the theoretical local permeability improvement for acid treatment with a slowly reacting flowing acid.[11]

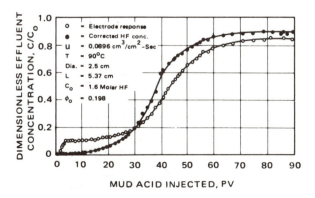

Fig. 8.5—Typical hydrofluoric acid effluent concentration profile.[13]

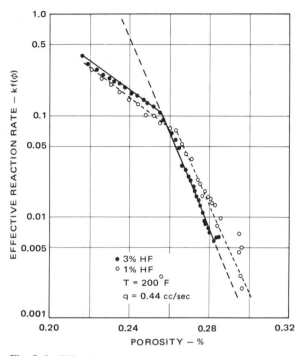

Fig. 8.6—Effective reaction rate vs porosity (Berea sandstone cores).[13]

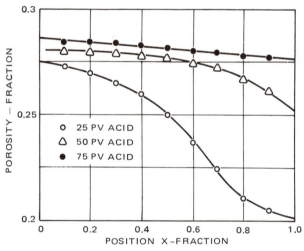

Fig. 8.7—Computed porosity distribution.[13]

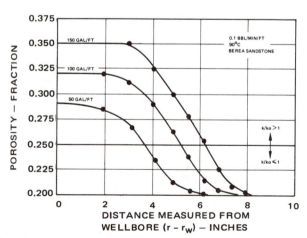

Fig. 8.8—Predicted porosity distribution for radial acid flow from 3-in. wellbore.[13]

function of porosity for a Berea sandstone core. The intersection of the two straight lines seen in this figure is interpreted to be the point at which the accessible clays were dissolved.

Once the reaction-rate coefficient is determined, finite-difference computational techniques can be used to solve the acid balance equation (Eq. 8.15) to obtain the acid concentration as a function of time and position for linear systems. The acid balance equation also can be expressed in radial or spherical coordinates and solved to compute the change in porosity as the acid treatment progresses.[13] The results of a numerical solution are shown in Fig. 8.7 for a linear system. Fig. 8.8 shows results for a radial system; note the progressive increase in the radius of the zone of porosity increase as acid is injected.

To understand acidization better in "real formations," Williams[15] conducted an experimental study of the response of several typical Gulf Coast sandstones. These studies showed that (1) because of differences in composition, the relationship between the effective reaction rate and porosity shown in Fig. 8.6 did not apply to all formations and (2) even when the compositions were similar, the responses can be very different, apparently because of the way clays were distributed in the rock.

The reaction-rate response curve from a Frio sandstone was typical of the sandstones studies and therefore was used to develop design curves for a "typical" formation. A sample design curve is shown in Fig. 8.9. Curves for temperatures ranging from 100 to 250°F and specific injection rates varying from 0.001 to 0.2 bbl/min/ft are presented in Chapter 9.

Ideally, one should conduct acid reaction rate tests to simulate what will occur in the formation to be treated. Realistically, this is seldom practical. The application of these curves to the design of an acid treatment is discussed in Chapter 9.

The Fogler-McCune Model

Fogler, McCune, and various coworkers[14, 16, 17] also have considered the problem of determining the depth of acid penetration during acidization. Their approach is somewhat

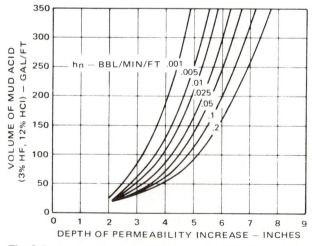

Fig. 8.9—Depth of permeability increase for 150°F formation temperature and 3-in. wellbore radius.[15]

different than that adopted by Williams and Whiteley[13] in that a steady state was not assumed. Instead, both the reaction of the acid with silica and the porosity changes attending acidization were neglected. Based on these assumptions, an analytical solution of the acid balance equations can be obtained if it is also assumed that the reaction rate of the acid with the rapidly dissolved minerals is characterized by a single-rate model. The acid balances solved are given by Eqs. 8.18 and 8.19.

$$\phi_o \frac{\partial \overline{c}}{\partial t} + u \frac{\partial \overline{c}}{\partial X} = -\widetilde{r}_{\text{ave}}, \quad \dots \dots \quad (8.18)$$

and

$$\frac{\partial}{\partial t}\left[(1 - \phi_o)c_m\right] = -\frac{\widetilde{r}_{\text{ave}}}{\widetilde{\beta}}, \quad \dots \dots \dots \quad (8.19)$$

where c_m is the lumped concentration of all minerals other than silica and $\widetilde{r}_{\text{ave}}$ is the average reaction rate of the acid with the dissolved minerals. The quantity $\widetilde{\beta}$ is an average stoichiometric coefficient relating the moles of rock dissolved per mole of acid reacted.

The reaction rate is assumed to have the form

$$-\widetilde{r}_{\text{ave}} = \xi_a' \, \overline{c} \, (c_m - c_{mi}), \quad \dots \dots \quad (8.20)$$

where ξ_a' is a reaction rate constant that includes the surface area in contact with acid. This is not the ξ_f' defined in Chapter 4 and therefore must be determined experimentally. c_{mi} is that part of the dissolvable minerals not accessible to acid attack.

The material balance represented by Eq. 8.18 is similar to that used by Williams and Whiteley (Eq. 8.15). Both equations assume that the rate is first order in HF concentration — a good approximation (see Chapter 4) and both recognize that the rate must be related in some way to the change in mineral content. In Eq. 8.15, the rate is assumed to be a function of the porosity, whereas in Eq. 8.20, it is assumed proportional to the portion of the accessible soluble minerals not yet dissolved.

Eqs. 8.18 through 8.20 can be integrated[16] to give

$$\frac{c_o}{c} = 1 + \exp\left[N_{\text{Da}}N_{AC}\left(1 + \frac{1}{N_{\text{CAP}}}\right)\frac{X}{L} - \frac{tu}{\phi_o L}\right]$$

$$- \exp\left[N_{\text{Da}}N_{\text{CAP}}\left(\frac{X}{L} - \frac{tu}{\phi_o L}\right)\right] \quad \dots \dots \quad (8.21)$$

The dimensionless groups are defined as

$$N_{\text{Da}} \text{ (Damkohler number)} = \frac{\xi_a'(c_{mo} - c_{mi})L}{u}$$

$$N_{\text{CAP}} \text{ (Acid capacity number)} =$$

$$\frac{\phi_o c_o}{\widetilde{\beta}(1 - \phi_o)(c_{mo} - c_{mi})},$$

where c_{mo} is the initial concentration of dissolvable minerals, c_{mi} is the part of the dissolvable minerals that is not accessible, and L is the length of the linear system.

Fig. 8.10 shows typical acid concentration profiles computed using Eq. 8.21. A reaction front is illustrated in this figure as the zone over which the acid concentration undergoes maximum change. This front is seen to develop rapidly

and then propagate down the system. The rate of frontal movement is related to the acid capacity number, N_{CAP}. Fig. 8.10 also demonstrates the effect of neglecting the reaction of the acid with silica. The acid concentration ratio rapidly approaches unity after the front passes. This reaction is not, however, negligible as shown in Fig. 8.5, where the stabilized acid concentration is about 90 percent of the inlet concentration in a 5.37-cm long core at 200°F. As the length of the linear system increases, this stabilized acid concentration decreases and the observed acid concentration deviates from that predicted by Eq. 8.21. This deviation becomes more pronounced at elevated temperatures as the reaction rate with silica increases.

The predictions obtained using either the Williams-Whiteley approach or that proposed by Fogler and McCune will not be entirely correct. The Williams-Whiteley model, which assumes a steady state, will be least accurate during the initial stages of acid injection, and the Fogler-McCune approach will be inaccurate at either long times or significant distances from the point of injection.

8.7 Model of Wormhole Growth in Carbonate Acidization

As discussed in Section 8.5, the theoretical models for matrix acidization are not satisfactory for carbonates. This problem apparently occurs because a few very large pores will govern the reaction process. To better understand acid reaction in carbonates, Nierode and Williams[18] simulated mathematically and studied experimentally the parameters governing wormhole growth.

To model acid reaction in a wormhole, we must know the injection rate into the formation, the number of wormholes created by acid reaction, and the rate of acid fluid loss from the wormholes to the formation. Normally, neither the number of wormholes nor the rate of acid fluid loss from the wormhole to the formation is known. To determine the possible reaction rate limit of the wormhole length, a theoretical model that predicts the acid concentration profile along a wormhole was developed by Nierode.[20] The expected wormhole length (if limited by acid reaction) can be approximated as the point where the acid concentration is one-tenth the injected concentration.

The procedures used to develop the model are analogous to those used to simulate acid reaction in a fracture (see Chap-

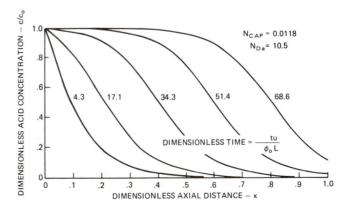

Fig. 8.10—Predicted acid concentration profiles in Phacocides sandstone.[16]

ters 5 through 7). The results for Nierode's analysis are plotted in Fig. 8.11. This figure relates the product of the Reynolds number for fluid loss and the Schmidt number ($N_{Re}* N_{Sc}$) for a given dimensionless acid concentration $\bar{c}/c_o$ to the dimensionless penetration distance ($4xLN_{Re}*/aN_{Re}$), where

$N_{Re}*$ = Reynolds number for fluid loss from a wormhole
= $2a\bar{v}_N\rho/\mu$,
N_{Re} = Reynolds number for flow along a wormhole = $2a\bar{v}_A°\rho/\mu$,
N_{Sc} = Schmidt number = $\mu/\rho D_e$,

and $\bar{v}_N$ is the average fluid loss velocity, a is the wormhole radius, and $\bar{v}_A°$ is the fluid velocity at the entrance to the wormhole.

Fig. 8.11 is comparable to Fig. 7.14, which was used to predict the acid penetration distance in an acid fracturing treatment. Note that as the Schmidt number decreases (that is, the effective mixing coefficient, D_e, increases), the distance along the wormhole that acid of a given concentration can penetrate will decrease. For example, when $N_{Re}* N_{Sc}$ is 3.0, acid with a concentration ratio of 0.25 would reach the end of the wormhole. If $N_{Re}* N_{Sc}$ is 0.3, a concentration of 0.25 would reach only about 20 percent of the distance along the wormhole.

To determine effective mixing coefficients for various acids and treating conditions, experiments were conducted in which HCl was pumped through a wormhole under carefully controlled conditions.[20] In these experiments, HCl was injected into an Indiana limestone core at a high velocity until a wormhole completely penetrated the core. After the wormhole was formed and enlarged to a diameter of about 0.25 in., the effluent acid concentration was measured at several flow rates at a temperature of 200°F. To calculate the effective mixing coefficient for these data, the model for acid reaction during flow through a circular channel (dis-

cussed in Chapter 4) was used. In the experiments of interest, the rate of acid reaction at the surface is not the controlling step in the over-all process, so this model is used in a form that relates the extent of acid reaction, $1 - \bar{c}/c_o$, to the dimensionless channel length, L_{cD}, for the assumption that reaction rate is very large, relative to diffusion rate, $P_c = \infty$ (shown in Fig. 8.12, see also Fig. 4.9). The dimensionless groups used in this figure are

$$P_c = \frac{a\xi_f'}{D_e},$$

$$R_c* = 1 - \frac{\bar{c}}{c_o},$$

$$L_{cD} = \frac{LD_e}{2\bar{v}_A a^2}. \quad \dots\dots\dots\dots (8.23)$$

Mass transfer measurements for HCl flow through the wormhole show that convective mixing controls the rate of acid transfer to the rock surface. The experimental data summarized in Fig. 8.13 consistently show mass transfer coefficients greater than the molecular diffusion coefficient. The peculiar shape of the curve is caused by the mixing mechanism that dominates at a particular Reynolds number.

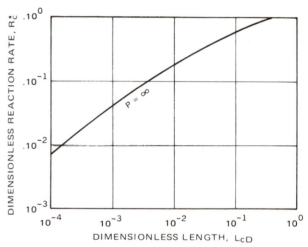

Fig. 8.12—Surface reaction rate in a circular tube (irreversible reaction).[19]

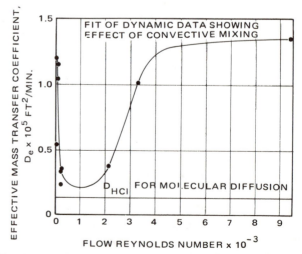

Fig. 8.13—The effective mixing coefficient for hydrochloric acid reaction in a wormhole (Indiana limestone).[20]

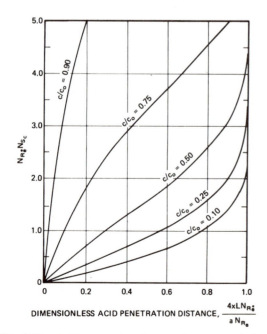

Fig. 8.11—Acid penetration distance along a wormhole.[20]

In the no-flow (static) limit, the observed intercept is about 10 times the molecular diffusion coefficient for HCl, D_{HCl}. Convection occurs even under static conditions because there is no orientation of the wormhole (with respect to gravity) that will be free from an adverse density gradient. As the Reynolds number increases to about 1,000, the amount of convection sharply declines to give a mass transfer coefficient near the molecular diffusion limit. This phenomenon can be ascribed to the laminar, dynamic flow condition with the short residence time in the wormhole not allowing full development of convection cells normal to the bulk flow direction. With Reynolds numbers above 1,000, turbulence effects again accelerate mixing to a level of about 10 times the molecular diffusion coefficient.

References

1. Dullien, F. A. L.: "Pore Structure Analysis," paper presented at the State-of-the-Art Summer Symposium, Div. of Ind. Eng. Chem., ACS, Washington, DC (June 1969).

2. Scheidegger, A. E.: *The Physics of Flow through Porous Media*, The Macmillan Co., New York (1960).

3. Purcell, W. R.: "Capillary Pressures — Their Measurement Using Mercury and the Calculation of Permeability Therefrom," *Trans.*, AIME (1949) **186**, 39-48.

4. Rose, W. D. and Bruce, W. A.: "Evaluation of Capillary Character in Petroleum Reservoir Rock," *Trans.*, AIME (1949) **186**, 127-142.

5. Burdine, N. T., Gournay, L. S., and Reichertz, P. P.: "Pore Size Distribution of Petroleum Reservoir Rocks," *Trans.*, AIME (1950) **189**, 195-204.

6. Klock, G. O.: "Pore Size Distributions as Measured by the Mercury Intrusion Method and Their Use in Predicting Permeability," PhD thesis, Oregon State U., Corvallis (1968).

7. Schechter, R. S. and Gidley, J. L.: "The Change in Pore Size Distribution from Surface Reactions in Porous Media," *AIChE Jour.* (May 1969) **15**, 339-350.

8. Guin, J. A.: "Chemically Induced Changes in Porous Media," PhD dissertation, U. of Texas, also Report No. UT 69-2, Texas Petroleum Research Committee, Austin (1969).

9. Guin, J. A., Silberberg, I. H., and Schechter, R. S.: "Chemically Induced Changes in Porous Media," *Ind. Eng. Chem. Fund.* (1971) **10**, 50-54.

10. Glover, M. C. and Guin, J. A.: "Permeability Changes in Dissolving Porous Media," *AIChE Jour.* (Nov. 1973) **19**, 1190-1195.

11. Sinex, W. E., Schechter, R. S., and Silberberg, J. H.: "Dissolution of a Porous Matrix by a Slowly Reacting Acid," *Ind. Eng. Chem. Fund.* (1972) **11**, 205-209.

12. Guin, J. A. and Schechter, R. S.: "Matrix Acidization with Highly Reactive Acids," *Soc. Pet. Eng. J.* (Sept. 1971) 390-398; *Trans.*, AIME, **251**.

13. Williams, B. B. and Whiteley, M. E.: "Hydrofluoric Acid Reaction With a Porous Sandstone," *Soc. Pet. Eng. J.* (Sept. 1971) 306-314; *Trans.*, AIME, **251**.

14. McCune, C. C., Fogler, J. S., Lund, K., Cunningham, J. R., and Ault, J. W.: "A New Model of the Physical and Chemical Changes in Sandstone During Acidizing," *Soc. Pet. Eng. J.* (Oct. 1975) 361-370.

15. Williams, B. B.: "Hydrofluoric Acid Reactions with Sandstone Formations," *J. Eng. Ind.* (Feb. 1975) 252-258; *Trans.*, ASME, **97**.

16. Lund, K., Fogler, H. S., and McCune, C. C.: "On Predicting the Flow and Reaction of HCl/HF Acid Mixtures in Porous Sandstone Cores," *Soc. Pet. Eng. J.* (Oct. 1976) 248-260; *Trans.*, AIME, **261**.

17. Lund, K.: "On the Acidization of Sandstone," PhD dissertation, U. of Michigan, Ann Arbor (1974).

18. Nierode, D. E. and Williams, B. B.: "Characteristics of Acid Reaction in Limestone Formations," *Soc. Pet. Eng. J.* (Dec. 1971) 406-418; *Trans.*, AIME, **251**.

19. Williams, B. B., Gidley, J. L., Guin, J. A., and Schechter, R. S.: "Characterization of Liquid/Solid Reactions — The Hydrochloric Acid/Calcium Carbonate Reaction," *Ind. Eng. Chem. Fund.* (Nov. 1970) **9**, 589-596.

20. Nierode, D. E.: Private communication, Esso Production Research Co., Houston.

Chapter 9

Matrix Acidizing of Sandstones

9.1 Introduction

This chapter discusses problems associated with contacting sandstone formations with mixtures of HF and HCl, gives procedures for designing treatments, and shows field results illustrating the more important principles. Optimal treatment design requires an understanding of the fundamentals of the chemical reactions involved and the corresponding physical processes that take place. Stoichiometry and reaction rate are discussed in Chapters 3 and 4 and fundamentals of matrix acidizing in Chapter 8. The reader should consult these chapters for important background information.

9.2 Description of a Sandstone Acidizing Treatment

An acidizing treatment for a sandstone formation normally will consist of sequentially injecting three fluids — a preflush, the hydrofluoric acid-hydrochloric acid mixture, and an afterflush. These fluids serve definite purposes.

The preflush is usually hydrochloric acid, ranging in concentration from 5 to 15 percent and containing a corrosion inhibitor and other additives as required. The preflush displaces water from the wellbore and connate water from the near-wellbore region, thereby minimizing direct contact between sodium and potassium ions in the formation brine and fluosilicate reaction products. Normally, this will eliminate redamaging the formation by precipitation of insoluble sodium or potassium fluosilicates (see Chapter 3, Section 3.5 for more details). The acid also reacts with calcite (calcium carbonate) or other calcareous material in the formation, thereby reducing, or eliminating, reaction between the hydrofluoric acid and calcite. The preflush avoids waste of the more expensive hydrofluoric acid and prevents the formation of calcium fluoride, which can precipitate from a spent HF-HCl mixture.

The *HF-HCl* mixture (usually 3-percent HF and 12-percent HCl) then is injected. The HF reacts with clays, sand, drilling mud, or cement filtrate to improve permeability near the wellbore. The HCl will not react and is present to keep the pH low, preventing precipitation of HF reaction products.

An *afterflush* is required to isolate the reacted HF from brine that may be used to flush the tubing and to restore water wettability to the formation and the insoluble-acid reaction products. Normally, one of three types of afterflush is used (1) for oil wells, either a hydrocarbon afterflush, such as diesel oil, or 15-percent HCl is used; (2) for water injection wells, hydrochloric acid is used; and (3) for gas wells, either acid or a gas (such as nitrogen or natural gas) is used. With a liquid afterflush chemicals usually are added to aid in removing treating fluids from the formation, restoring water wettability to formation solids and precipitated acid reaction products, and prevention of emulsion formation. A glycol ether mutual solvent has been shown to be useful for this purpose.[1,2] When a gas is used as an afterflush, cleanup additives are added to the HF-HCl stage of the treatment.

9.3 Mechanism of Acid Attack

Damage Induced by Acid

Studies of the mechanism of acid attack in sandstone formations were first published by Smith *et al.*[3] In their studies, core plugs were acidized under carefully controlled conditions. The permeability of the core was studied during injection of HF-HCl acid mixtures to monitor changes that occurred for various acid concentrations, applied pressure gradients (flow rates), and for different sandstone formations. A plot of permeability change as a function of the amount of acid injected was subsequently called an Acid Response Curve[4] (Fig. 9.1).

These studies were the first to show that core permeability declines on initial contact with mixtures of HCL and HF acids. Upon continued injection of the HF-HCl mixture permeability increases, as shown in Fig. 9.1. Smith *et al.* reasoned that the initial permeability reduction was caused by the partial disintegration of the sandstone matrix and the downstream migration of fines that plug flow channels in the core. Continued exposure of the fines to unspent HF was thought eventually to result in their dissolution. Therefore, the subsequent permeability increase was thought to come from clearing the pore channels plugged by fines and the enlargement of other pore channels by the acid. As discussed in Chapter 3, Labrid[5,6] has recently taken extensive equilibrium data for the sandstone-HF reaction and has proposed that the decrease in permeability observed during acidization could be caused by the precipitation of orthosilicic acid or other reaction products. The same sort of permeability decrease may be seen when only HCl acid is injected into a sandstone (no precipitates are expected). In this case, plugging could be caused by either, or both, of these mechanisms.

The Effect of HF Concentration on Core Response to HF-HCl

Fig. 9.1 illustrates the effect of hydrofluoric acid concentration on the response of Berea sandstone to HF-HCl mixtures. This figure shows that higher HF concentrations give a greater initial permeability decrease, but that a smaller volume of acid will achieve a given permeability increase. The Berea sandstone used in these tests is a relatively homogeneous sandstone that contains less clay (the most reactive component) than many sandstone formations; therefore, the response may be different in formation sands.

These data indicate that from 50 to 100 PV of a 3-percent HF and 12-percent HCl mixture have to be injected to achieve a significant permeability increase. Using this acid response curve and data presented in Table 2.1 to estimate the pore volume in a 1-ft circle surrounding the wellbore, we predict that an acid volume of 220 to 440 gal of acid mixture/ft of interval may be required to treat a zone of 1-ft radius around the wellbore effectively. It is important to recognize that this volume of acid may destroy the consolidating materials in the rock and allow permeability to decrease because of compaction (see Figs. 9.4 and 9.5).

The Effect of Pressure Gradient on Core Response to HF-HCl

Fig. 9.2 shows that as the acid flow rate through the Berea core is increased (pressure gradient is increased), the initial permeability decline increases. Also, greater quantities of acid are required to achieve a given permeability increase.

The increased permeability decline may be caused by an increase in the quantity of fines released because of the increased drag forces at high flow rates. Larger volumes of

acid are probably required to achieve a given permeability increase because all the HF was not reacted while the acid resided in the core when injected at the higher flow rates. Since effluent acid concentrations were not reported, this effect cannot be verified.

The Effect of Matrix Composition on Core Response to HF-HCl

The mineralogical composition of the sandstone matrix has a substantial effect on a formation's response to hydrofluoric acid. Berea sandstone, a relatively clean quartzitic material usually containing about 6-percent clay, was used for the tests depicted in Figs. 9.1 and 9.2. Because of its low clay content, this material shows only a slight reduction in permeability on initial acid contact. Fig. 9.3 illustrates the correspondingly greater reduction in permeability on initial acid contact with other sandstones. Core C, which contains more quartzitic fines than clays, shows the greatest reduction in permeability of the formations tested. Quartzitic fines are slower to react with mud acid than are clay minerals and, once loosened from the matrix, are more effective in plugging pore channels. The Core C results indicate that formations containing quartzitic fines require more acid to attain a given permeability increase than do formations containing mostly clay minerals.

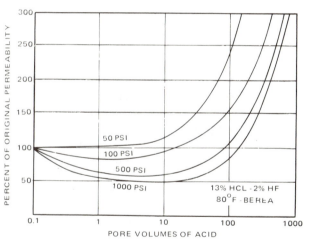

Fig. 9.2—Effect of acid flow rate on Berea core response to HF-HCl.[3]

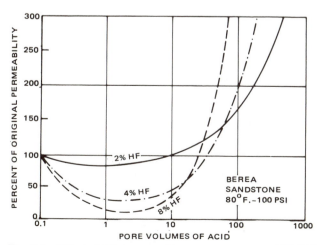

Fig. 9.1—Effect of HF concentration on core response to HF-HCl mixtures.[3]

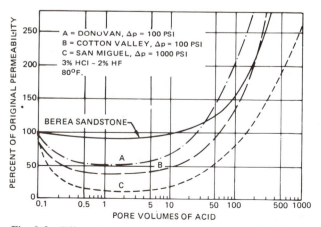

Fig. 9.3—Effect of core composition on response to HF-HCl.[3]

Effect of HF-HCl Reaction on Core Mechanical Properties

In an effort to remove damage completely, whether originally present or induced by the acid, one may decide to try a larger acid volume. Aside from the additional cost, there is a physical limitation on the quantity of acid the formation can tolerate without becoming unconsolidated. Recall that the acid is dissolving the cementing material; therefore, as acid is injected, the formation progressively becomes weaker until it finally disintegrates.

The effect of elevated temperatures, pressures, and confining load on the response to HF-HCl mixtures was studied by Farley *et al*.[7] Fig. 9.4, taken from this study, illustrates what happens to the compressive strength of a core as increased volumes of acid are injected. These data show that as the volume of acid injected is increased, the uniaxial compressive strength decreases until the sandstone is finally unconsolidated. Note that the compressive strength decrease correlates closely with the total dissolving power of acid injected. For example, a compressive strength of 500 to 600 psi was obtained after injection of an equivalent of about 18 gal/ft of 8-percent HF, 30 gal/ft of 5-percent HF, and 75 gal/ft of 2.5-percent HF.

If simulated overburden stresses are imposed on a core during acidization, a point is soon reached where the compressive strength of the core is inadequate to support the load and the core recompacts to a lower porosity and permeability. This effect is shown in Fig. 9.5. Note that the progressive improvement in permeability with acid throughput reverses once sufficient acid has been injected to remove consolidating material from the sand. This reversal begins for the sandstones studied here after injecting about 18 gal/ft of 8-percent HF or 40 to 60 gal/ft of 5-percent HF, but has not yet begun after injection of 120 gal/ft of 2.5-percent HF. The reversal corresponds roughly to reducing the core's compressive strength to about 200 to 500 psi (see Fig. 9.4).

The amount of acid required to remove formation consolidation varies with many factors — the formation's initial compressive strength, its depth, its mineralogy, and how rapidly cementation is dissolved by the acid. Field experience usually determines the appropriate acid volume to be used, although laboratory tests can estimate the treatment volume for initial treatments in an area.

9.4   Prediction of Radius of Acid Reaction

To predict the radius of acid attack, and thus the quantity of acid required for a treatment, we must consider reaction kinetics and changes in formation properties caused by acid reaction. The kinetics of HF sandstone reactions are considered in Chapter 4. Analytical approaches to predict changes caused by acidization are considered in Chapter 8.

The approach we use in designing sandstone acid treatments was developed by Williams and Whiteley.[8,9] This technique (described in detail in Chapter 8) couples a mathematical description of acid reaction with data taken on formation core material. The resulting design curves are shown in Figs. 9.6 through 9.9. These figures are for temperatures ranging from 100 to 250°F and injection rates of 0.001 to 0.2 bbl/min/ft of formation to be treated. The curves were developed for 3-percent HF/12-percent HCl, but the effect of other acid concentrations can be estimated by converting to the equivalent volume of 3-percent HF on a dissolving power basis.

When presenting these curves, Williams and Whiteley pointed out that acid reaction rate tests should be conducted to determine the appropriate reaction rate coefficient for each formation of interest before attempting the quantitative design of acid treatments. The impracticality of this sugges-

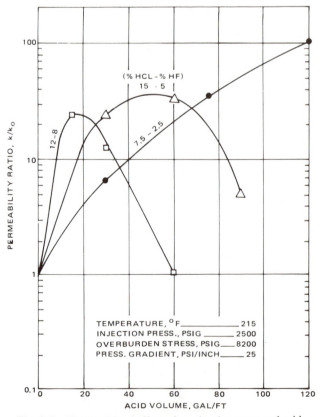

Fig. 9.5—The combined effect of overburden stress and acid throughput on permeability.[7]

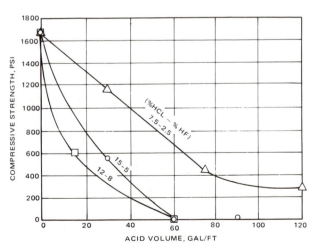

Fig. 9.4—Effect of acid throughput on formation compressive strength.[7]

tion was recognized, however, and average response curves were proposed to obtain a reasonable estimate of the required acid volume without detailed experimental data. Use of the curves is illustrated in a subsequent section.

Other authors have proposed different methods of computing the depth of acid penetration. For example, Gatewood et al.[10] proposed that the acid penetration distance be predicted by assuming instantaneous reaction of HF acid with clays in the formation and that all clays are contacted with acid. Measured reaction rates with silica sands were included as a time-dependent reaction. Hence, knowing the clay content of the formation and the dissolving power of the acid, the penetration depth can be directly related to the volume of acid injected. This approach is useful for estimating the depth of acid penetration and often gives results consistent with Figs. 9.6 through 9.9, but its use is limited by a lack of design curves in the study. Inherent in this application is the assumption that the acid will effectively contact all clay components in the sandstone or that the fraction of clays to be contacted is known.

McCune et al.[11] proposed a method of calculating the depth of acid penetration, but their concept ignores the silica reaction; it employs a slightly different kinetic approach but exactly the same basic equation as Williams and Whiteley (both procedures are discussed in Chapter 8). Although McCune et al.'s approach results in design curves that simplify acid-volume calculations, the curves imply a relatively sharp reaction front (because the silica reaction is ignored) and commonly predict penetration distances larger than those observed in practice.

9.5 Productivity After Sandstone Acidizing

In Chapter 2 we indicated that a matrix acid treatment in a sandstone formation should remove flow restrictions near the wellbore and allow the well to produce at an undamaged rate. However, the production rate after treatment is sometimes lower than expected. Flow restrictions after treatment are believed to occur most often because of precipitation of acid reaction products, a change in the wettability of the rock from water wet to oil wet, or the formation of an emulsion between the reacted acid and formation oil.

Precipitated reaction products can plug formation pores

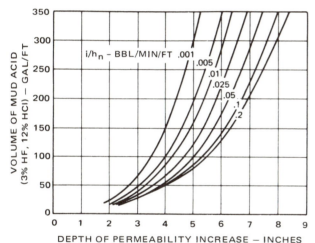

Fig. 9.6—Depth of permeability increase for 100°F formation temperature and 3-in. wellbore radius.[9]

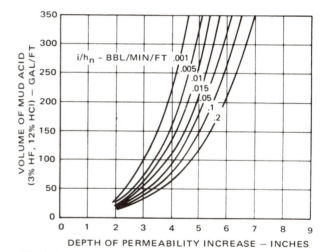

Fig. 9.8—Depth of permeability increase for 200°F formation temperature and 3-in. wellbore radius.[9]

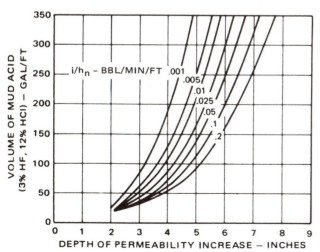

Fig. 9.7—Depth of permeability increase for 150°F formation temperature and 3-in. wellbore radius.[9]

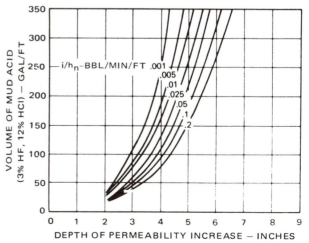

Fig. 9.9—Depth of permeability increase for 250°F formation temperature and 3-in. wellbore radius.[9]

and restrict flow in the same manner as clay particles from drilling mud. Precipitation can occur as reaction proceeds (proposed by Labrid[5]) and as a result of contact between the reacted acid and formation brine, or a brine used as an afterflush. The following steps will minimize the chance for redamage to the formation by precipitation: (1) an acid preflush should be used, (2) an afterflush compatible with the reacted acid should be used, and (3) the acid should be produced back out of the formation within a few hours of treatment completion.

Both oil wetting of the formation matrix and fine particles released by acid reaction can potentially reduce productivity. The formation solids may be oil wet because the corrosion inhibitor or other additives adsorb strongly on clay particles or clean silica surfaces. Oil wetting of these surfaces can reduce the relative permeability to oil.

The effect of a wettability change on productivity can be illustrated schematically with Fig. 9.10, an example relative-permeability curve for a water-wet rock. Plotted here is the relative permeability, as a percent of the absolute permeability, for the wetting phase, k_{rw}, and the nonwetting phase, k_{ro}. Assume, for example, that the water saturation is 40 percent. Then, from Fig. 9.10, $k_{ro} = 60$ percent and $k_{rw} = 2$ percent. That is, the permeability to oil is 60 percent of the absolute rock permeability and the permeability to water is 2 percent. For illustrative purposes, if the rock is made totally oil wet, the permeability relationship is essentially reversed. (In actual practice an exact interchange will seldom, if ever, occur.) Oil (now the wetting phase) will have a relative permeability of 2 percent and water a relative permeability of 60 percent. From Fig. 2.2 we see that a change of permeability of this magnitude, even if restricted to the first 6 to 12 in. around the wellbore, may give a two- to threefold reduction in productivity when compared with the undamaged formation productivity.

Productivity impairment also can occur if an emulsion is formed between the reservoir oil and the reacted acid. Often these emulsions are stabilized by fines that are partially oil wet and partially water wet. Impairment is greatest when the drop size in the emulsion is large enough so that the drops cannot easily pass through restrictions in the porous rock matrix. Early in this century emulsions stabilized by fine particles were recognized by the scientific community.[12-14] They are often described as solids-stabilized or mechanical emulsions. Gidley[1] described the role of such emulsions in the stimulation of sandstone formations by mixtures of hydrochloric-hydrofluoric acid.

Stabilization of emulsions by solids (illustrated in Fig. 9.11) depends critically on the wettability of the fines produced by acid attack. The most effective stabilizers are fines with mixed wettability and of suitable particle size — generally less than 2 microns in diameter.[14, 15] Mixed wettability is required to allow the particles to become positioned at the oil-water interface, while the small particle-size requirement is largely a geometric need related to the size of the emulsion droplets. Clearly, the solid particles must be much smaller in diameter than the internal phase emulsion droplet; a cross-section of one is shown in Fig. 9.11. Also, pH, oil composition, and other factors can influence emulsion formation and stabilization.

When emulsification results during stimulation, the emulsified liquid can exhibit an extremely high viscosity and very poor flow characteristics. Therefore, the emulsion impedes the flow of spent mud acid from the formation to the wellbore and reduces the effectiveness of the stimulation treatment by occupying pore spaces near the wellbore. If the emulsion is incompletely stabilized, this productivity impairment may be only temporary. Stable emulsions, however, may have a lasting detrimental effect on the productivity of a well. The effect of additives and pH on solid wettability is complex and can only be studied in laboratory tests conducted with all components present. Test proce-

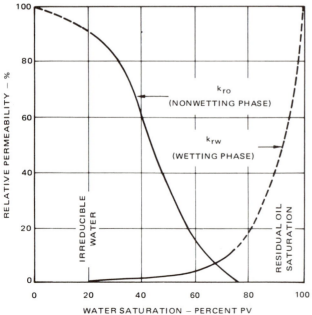

Fig. 9.10—Example relative permeability curves.

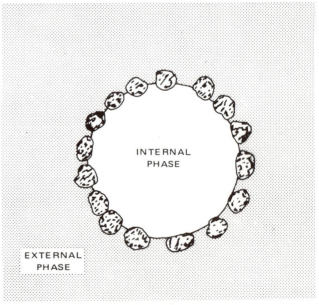

Fig. 9.11—To stabilize an emulsion solid particle, diameter must be small compared with emulsion droplet.[15]

dures are discussed in API RP 42.

Productivity restrictions caused by oil wetting of formation and fines often can be removed by making the solids water wet. This may be done with surface-active materials *provided they contact the solid surface and replace the agent making the surface oil wet. Unfortunately, most surface-active materials alone cannot cause the oil wetting agent to be desorbed.* Some chemicals (such as the low molecular weight glycol ethers) can strip the oil wetting surfactant from the surface and leave it water wet. Ethylene glycol monobutyl ether (EGMBE) is often preferred for this application. When used at a concentration of 10 percent by volume in the afterflush, this material is most effective.[13] Low molecular weight alcohols are sometimes used for this purpose, although they appear less effective than the glycol ethers. A statistical comparison between treatments using EGMBE and those employing other afterflush additives is shown in Table 9.1.[16] Note that both the success ratio and productivity improved substantially when EGMBE was used. Additional discussion of mutual solvents is included in Chapter 11, Section 11.4. These results should not be assumed to imply that mutual solvents can, or should, be used indiscriminately. Compatibility tests should be performed with acid, formation fluids, and formation solids to assure damage will not occur.

9.6 Design Procedure for Sandstone Acidizing Treatments

The design of an acid treatment should specify not only the volumes and types of fluids to be injected, but also the maximum permissible injection rate and treating pressure (to avoid fracturing the formation). A systematic method for designing a sandstone acidizing treatment is outlined below.

Design Procedure

Step 1. Determine the fracture gradient for the well or field of interest. The best data are obtained from the instantaneous shut-in pressure measured during or immediately after a fracturing treatment. If no recent data are available, the gradient can be estimated with the approximate relation given as Eq. 7.2 (see Chapter 7, Section 7.2 for a discussion of its use).

$$g_f = \alpha + (\text{overburden gradient} - \alpha)$$
$$\frac{\text{reservoir pressure}}{\text{depth}} \quad \ldots \ldots \ldots \ldots \ldots (7.2)$$

Step 2. Predict the maximum possible injection rate without fracturing.

$$i_{max} =$$
$$\frac{4.917 \times 10^{-6} k_{av} h_n (g_f \times \text{depth} - \text{reservoir pressure})}{\mu \ln(r_e/r_w)} .$$
$$\ldots \ldots \ldots \ldots \ldots \ldots \ldots \ldots \ldots \ldots (9.1)$$

To use this equation, the permeability is expressed in millidarcies, viscosity in centipoise, pressure in psi, while drainage radius, wellbore radius (r_w), and net thickness of the sand are expressed in feet. The computed injection rate is in barrels per minute.

TABLE 9.1—SUMMARY OF OILWELL ACID TREATMENT RESULTS, LOUISIANA AND EASTWARD[16]

Type of Treatment	Number of Procedures	Percent Success	Average Production Increase (BOPD/well)
HF-HCl; EGMBE	23	65	121.0
HF-HCl; no EGMBE	12	25	18.3
HCl; EGMBE	5	60	35.2
HCl; no EGMBE	14	22	18.8

To avoid fracturing the formation, the injection rate must clearly be lower than i_{max}. As a practical matter, a rate 10 percent lower is suggested. It may not be possible to attain even this reduced rate without fracturing the formation because the injection rate is initially controlled by damage zone permeability rather than by formation permeability. Note that k_{av} is the average permeability, including both the damage zone and formation matrix permeabilities, not the matrix permeability far from the wellbore. Acid viscosities used in Eq. 9.1 can be obtained from Fig. 9.12.

Step 3. Predict maximum surface pressure, ignoring friction down tubing, for which fluids can be injected without fracturing the formation:

$$p_{max} = (g_f - \text{acid hydrostatic gradient}) \text{ depth}. \quad \ldots (9.2)$$

Data for the acid hydrostatic gradient can be obtained from Fig. 9.13.

During the treatment, surface pressure should be limited to pressures below this maximum. *If injection is not possible without exceeding this pressure,* it may be exceeded initially, but as soon as injection is started the rate should be reduced to maintain surface pressure lower than p_{max}.

Step 4. Determine the volume of HCl-HF solution to use (see design charts given as Figs. 9.6 through 9.9).

● Divide the maximum injection rate determined in Step

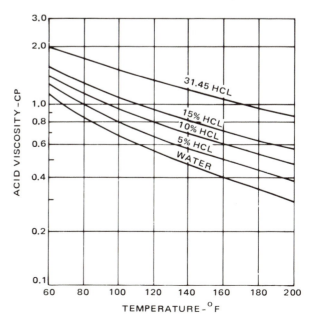

Fig. 9.12—Surface reaction rate in a circular tube (irreversible reaction).[19]

2 by the perforated interval thickness to obtain the specific injection rate, in barrels/minute per foot of perforated interval.

- Estimate the radius of the damage zone. In the absence of well test data, the authors suggest that low-permeability (less than 5 md) formations have a damage zone thickness estimated at 3 in.; more permeable formations may be regarded as having a damage zone of 6 in. or more.
- Determine formation temperature (in °F).
- Choose the chart from Figs. 9.6 through 9.9 nearest the formation temperature of concern and, using the assumed damage zone thickness and specific injection rate, read the volume of mud acid (3-percent HF, 12-percent HCl) required to obtain a permeability increase to the desired radius. If the mud acid to be used contains more than 3 percent HF, reduce the volume read from Figs. 9.6 through 9.9 by multiplying by the ratio 3/(HF concentration). If the wellbore radius (r_w) is not 3 in., use Eq. 9.3 to compute the required volume.

$$V = V_o \frac{(r_w + \Delta r_{acid})^2 - r_w^2}{(3 + \Delta r_{acid})^2 - 9}, \quad \dots \dots \dots \dots \text{(9.3)}$$

where V_o is the volume of mud acid read from Figs. 9.6 through 9.9 required when the wellbore radius is 3 in., and Δr_{acid} is the depth of permeability increase into the formation measured in inches. In wells with long open intervals, economics may limit the total acid volume that can be used. For these wells, Figs. 9.6 through 9.9 can be used to predict the radius treated with a given acid volume.

Step 5. Specify the treatment. The acid treatment is specified as follows.

Preflush. Normally, inject 50 gal of regular acid per foot of perforated interval. The preflush is intended (1) to remove calcite from the near-wellbore region before it comes in contact with HF and (2) to form a barrier between the HF-HCl mixture and the formation brine.

Hydrofluoric Acid. Inject the volume of mud acid determined from the design chart.

Afterflush. In oil wells, inject a volume of afterflush of diesel oil or hydrochloric acid equal to the mud acid volume.

(Note, as discussed earlier, field studies show that often an afterflush composed of 90-percent diesel oil and 10-percent ethylene glycol monobutyl ether improves results on the average.) The afterflush isolates spent acid from fluids used to displace the treating fluids into the formation. The diesel oil-EGMBE mixture also removes water from the formation near the wellbore, improves relative permeability to oil, and leaves the formation strongly water wet.

In gas wells and water injection wells, the afterflush is normally 15-percent HCl. Addition of 10 percent by volume EGMBE is again recommended. Diesel oil is not to be used in either gas or water-injection wells because of its adverse effect on the relative permeability to either fluid.

No soaking time is required for this treatment to be effective. As soon as possible after injection is complete, the well should be put back on production. Prolonged waiting periods, with spent acid in the formation, reduce the effectiveness of the treatment.

Example Design Calculation

To illustrate the design procedure, we will consider the design of an acid mutual solvent treatment for an oil well with the following characteristics: (1) formation depth = 5,000 ft, (2) perforated interval = 10 ft = h_n, (3) k_{av} = 50 md (average permeability of the formation, including the damaged zone, before acid treatment), (4) temperature = 150°F, (5) μ_o = 1 cp at reservoir conditions, (6) μ = 0.78 cp (viscosity of 15-percent HCl at 150°F from Fig. 9.12), (7) fracture gradient = 0.7 psi/ft (at initial pressure of 2,000 psi), (8) current reservoir pressure = 1,000 psi, (9) overburden gradient = 1.0 psi/ft, (10) wellbore radius = 3 in., and (11) drainage radius = 660 ft.

Step 1. Fracture gradient is not known for *current* reservoir pressure, so this must be estimated using Eq. 7.2, the overburden gradient, and the *initial* reservoir pressure and fracture gradient.

$$0.7 = \alpha + (1-\alpha) \frac{2,000 \text{ psi}}{5,000 \text{ ft}},$$

solving for α, we find

$$\alpha = 0.5,$$

and the current fracture gradient is estimated to be

$$g_f = 0.5 + 0.5 \times \frac{1,000 \text{ psi}}{5,000 \text{ ft}},$$

$$g_f = 0.6 \text{ psi/ft}.$$

Step 2. The maximum injection rate can now be estimated from Eq. 9.1,

$$i_{max} =$$

$$\frac{4.917 \times 10^{-6} (50 \text{md})(10 \text{ft})(5,000 \text{ft} \times 0.6 \text{ psi/ft} - 1,000 \text{psi})}{(0.78 \text{ cp}) \ln(660/0.25)}$$

$$i_{max} = 0.80 \text{ bbl/min}$$

$i/h = 0.080$ bbl/min/ft [in practice we would prefer to limit the rate to 0.72 bbl/min (0.9 × 0.80 bbl/min)].

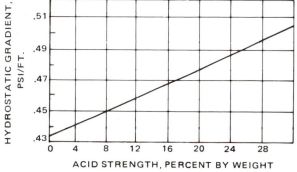

Fig. 9.13—Pressure gradient imposed by hydrochloric acid in wellbore.

Step 3. Maximum surface pressure should not exceed that given by Eq. 9.2.

$$p_{max} = (0.6 - 0.47) \text{ psi/ft} \times 5,000 \text{ ft,}$$

where 0.47 psi/ft is the hydrostatic gradient of 15-percent HCl (Fig. 9.13),

$$p_{max} = 650 \text{ psi.}$$

Step 4. Acid volume required to give 6-in. depth of permeability change (obtained from Fig. 9.7) is

$$\text{Vol} = 220 \text{ gal/ft of perforated interval.}$$

Step 5. Job design is as follows.

Preflush:	500 gal of 15-percent HCl
Acid:	2,200 gal of 3-percent HF-12-percent HCl
Afterflush:	2,200 gal of 90-percent diesel oil-10-percent EGMBE

Displace all fluids into the formation at a surface pressure less than 650 psi in the sequence shown above. Stop the displacement when the displacing fluid reaches the top perforation.

Corrosion inhibitors that are effective in the presence of EGMBE should be selected to provide the required protection, taking into consideration type of tubular goods, temperature, and maximum acid-pipe contact time. The selection of corrosion inhibitors and other additives is discussed in Chapter 11.

9.7 Common Mistakes in Application of Acid Treatments

Although field experience with sandstone acid treatments has generally been good, when poor response is observed it may be caused by one of the following factors.

1. *Use of acids containing no hydrogen fluoride (HF).* Even though certain sandstone formations can be moderately stimulated with hydrochloric acid alone, stimulation of highly damaged formations or formations containing high concentrations of clay minerals normally requires an acid mixture containing HF. The trade names of some products can be misleading as they have names that imply mud cleanout but contain no HF.

2. *Lack of a hydrochloric acid preflush.* A preflush is required to eliminate mixing of salt water and mud acid. This mixing is detrimental because it leads to the formation of insoluble fluosilicate salts.

3. *Inadequate mud acid volume.* Some treatments are conducted with as little as 10 gal of mud acid per foot of formation. Even though the treatments may occasionally succeed if damage is extremely shallow or confined to the perforations, much higher success ratios can be assured with mud acid volumes of at least 125 gal per foot of perforated interval.[19] Substantially greater volumes of mud acid may be required in a formation that is highly permeable, extremely shaly, or extensively damaged.

4. *Lack of immediate cleanup.* Even with the outlined technique, detrimental reaction products will be formed if the mud acid is allowed to remain in the formation for an extended time. Some treatments on rather shallow formations have been performed satisfactorily with the mud acid

remaining in the formation overnight, but waiting longer than a day should be avoided. Generally, the well should be put on production as soon as possible after treatment — long shut-in times should be avoided, especially in high temperature formations. In some geologically young formations, Templeton *et al.*[18] recently contended that production rate should initially be restricted and slowly increased to prevent redamage by fines movement.

5. *Use of diesel oil in gas and water-injection wells.* In some cases, the injection of diesel oil (as an afterflush) in a gas or water-injection well reduces the gas or water relative permeability. This can reduce the cleanup rate and, apparently, in some reservoirs permanently reduces well productivity.

6. *Fracturing the formation during treatment.* Successful sandstone acidizing requires uniform invasion of the formation for damage removal near the wellbore.[16] Normally, little benefit is derived from a treatment that fractures the formation with mud acid. Mud acid is incapable of etching the formation enough to provide a conductive fracture. If a pressure exceeding p_{max} (Eq. 9.2) must be used to gain fluid entry, injection pressure should be reduced below p_{max} as soon as injectivity is established.

7. *No mutual solvent with mud acid treatment.* Field data show that the use of a mutual solvent (such as ethylene glycol monobutyl ether) in mud acid treatments will often improve both the frequency of success and well productivity of acid treatments. (See Table 9.1.) Caution: Before using a mutual solvent or any other additive, tests should be conducted to assure compatability with formation fluids and solids in the presence of acid and acid reaction products.

8. *Treating undamaged formation.* Mud acid is capable only of removing damage near the wellbore. Because of its very shallow penetration depth (see Figs. 9.6 through 9.9) it is incapable of providing reservoir stimulation. If a well that has low productivity is undamaged, a mud acid treatment will provide no productivity improvement and may, if improperly conducted, reduce productivity.

9.8 Future Trends in Sandstone Acidizing

Because the depth of permeability alteration is severely limited in HF-HCl treatments much research has been devoted to finding ways to extend the action of acid deeper into the formation. One new technique now undergoing field evaluation generates the HF acid in situ by injecting methyl formate and ammonium fluoride.[18-20] Methyl formate hydrolyzes in the presence of formation water to produce formic acid, which then reacts with ammonium fluoride to yield hydrogen fluoride. The reactions operable are given as Eqs. 9.5 and 9.6.

$$HCOOCH_3 + H_2O \rightleftarrows HCOOH + CH_3OH , \quad \ldots\ldots (9.5)$$

Methyl Formate	Water	Formic Acid	Methyl Alcohol

$$HCOOH + NH_4F \rightleftarrows NH_4^+ + HCOO^- + HF. \quad \ldots\ldots (9.6)$$

Formic Acid	Ammonium Fluoride	Ammonium Ion	Formate Ion	Hydrofluoric Acid

This acid system is retarded because of the slow rate of hydrolysis of methyl formate to create formic acid. Typical hydrolysis rate determinations reported by Templeton *et al.*[18] (Fig. 9.14) show that 90 percent hydrolysis occurs in the presence of Wyoming bentonite after about 6 hours at 158°F, 18 hours at 140°F, and 30 hours at 122°F.

To date, field results with this sytem have been reported for only one area where it was compared with conventional mud acid treatments — the South Pass Block 24 and Block 27 fields.[18] Self-generating mud acid treatments in the field are reported to involve injection of the following chemicals: (1) a xylene preflush, (2) a 10-percent HCL preflush, (3) a 7½-percent HCL/1½-percent HF preflush, (4) a 3-percent ammonium chloride spacer, and (5) the self-generating mud acid. Fluid volumes pumped and the composition of the self-generating mud acid were not specified.

After chemical injection, the wells were shut in long enough to allow complete acid reaction. They were then brought on production slowly over a period of *several weeks* to provide minimal disturbance to formation fines and gravel packs.

Fig. 9.15 is a set of oil production curves indicating early experience with this treatment in the South Pass Block 24 and 27 fields (results from 102 conventional and 28 self-generating mud acid treatments are included). A range of results is shown to include conventional treatments that involved both regular mud acid (12-percent HCl/3-percent HF) and half-strength mud acid. The upper band is believed to result from the use of a lower acid strength, low injection rate, and slow return to production after treatment.

The self-generating mud-acid results show an increasing production for the first 2 or 3 months (because of the slow increase in choke size) and an almost constant rate thereafter. Templeton *et al.* report this higher sustained rate is the main advantage of this acid system. They conclude that "after 7 months the self-generating acid is better than their best conventional treatments by almost a factor of two. The added cost of the treatment is rapidly recovered by this sustained higher productivity."

At this time we are not in a position to know when this process should be used in preference to the conventional treatment described previously since it was licensed recently for use throughout industry. Once its use becomes more widespread, a better over-all evaluation of the process should be possible.

References

1. Gidley, J. L.: "Stimulation of Sandstone Formations With the Acid-Mutual Solvent Method," *J. Pet. Tech.* (May 1971) 551-558.

2. Hall, B. E.: "The Effect of Mutual Solvents in Sandstone Acidizing," *J. Pet. Tech.* (Dec. 1975) 1439-1442.

3. Smith, C. F. and Hendrickson, A. R.: "Hydrofluoric Acid Stimulation of Sandstone Reservoirs," *J. Pet. Tech.* (Feb. 1965) 215-222; *Trans.*, AIME, **234**.

4. Smith, C. F., Ross, W. M., and Hendrickson, A. R.: "Hydrofluoric Acid Stimulation — Developments for Field Application," paper SPE 1284 presented at the SPE-AIME 40th Annual Fall Meeting, Denver, Oct. 3-6, 1965.

5. Labrid, J. C.: "Stimulation Chemique — Étude Théorique et Expérimentale des Équilibres Chimiques décrivant l'Attaque Fluorhydrique d'un Grès Argileux," *Revue* (Oct. 1971) L'Institut Français du Petrole, **26**, 855.

6. Labrid, J. C.: "Thermodynamics and Kinetic Aspects of Argillaceous Sandstone Acidizing," *Soc. Pet. Eng. J.* (April 1975) 117-128.

7. Farley, J. T., Miller, B. M., and Schoettle, W.: "De-

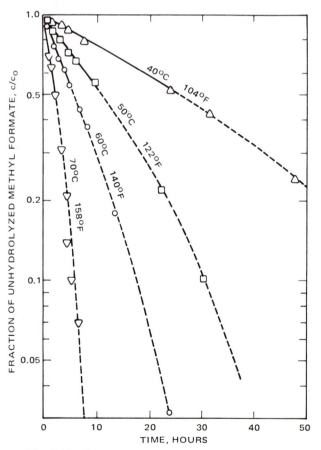

Fig. 9.14—Hydrolysis of methyl formate in 2.0-percent, self-generating mud acid in presence of Wyoming bentonite.[18]

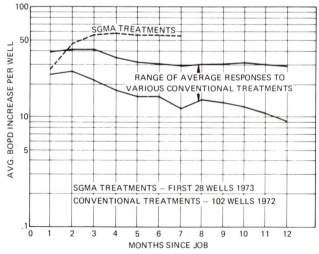

Fig. 9.15—Comparison of acid stimulation techniques for East Bay fields.[18]

sign Criteria for Matrix Stimulation with Hydrochloric-Hydrofluoric Acid," *J. Pet. Tech.* (April 1970) 433-440.

8. Williams, B. B. and Whiteley, M. E.: "Hydrofluoric Acid Reaction With a Porous Sandstone," *Soc. Pet. Eng. J.* (Sept. 1971) 306-314; *Trans.*, AIME, **251**.

9. Williams, B. B.: "Hydrofluoric Acid Reactions with Sandstone Formations," *J. Eng. Ind.* (Feb. 1975) ASME, 252-258.

10. Gatewood, J. R., Hall, B. E., Roberts, L. D., and Lasater, R. M.: "Predicting Results of Sandstone Acidizing," *J. Pet. Tech.* (June 1970) 693-700.

11. McCune, C. C., Fogler, J. S., Lund, K., Cunningham, J. R., and Ault, J. W.: "A New Model of the Physical and Chemical Changes in Sandstone During Acidizing," *Soc. Pet. Eng. J.* (Oct. 1975) 361-370.

12. Pickering, S. U.: "Emulsions," *J. Chem. Soc.* (1907) 2001.

13. Finkle, P., Draper, H. D., and Hildebrand, J. H.: "The Theory of Emulsification," *Journal,* ACS (1923) **45**, 2780.

14. Schulman, J. H. and Leja, J.: "Control of Contact Angles at the Oil-Water-Solid Interfaces," *Trans.*, Faraday Society (June 1954) **50**, No. 374, 598.

15. Adamson, A. W.: *Physical Chemistry of Surfaces,* Wiley-Interscience, New York (1960).

16. Gidley, J. L., Ryan, J. C., and Mayhill, T. D.: "Study of the Field Application of Sandstone Acidizing," *J. Pet. Tech.* (Nov. 1976) 1289-1294.

17. Sutton, G. D. and Lasater, R. M.: "Aspects of Acid Additive Selection in Sandstone Acidizing," paper SPE 4114 presented at the SPE-AIME 47th Annual Fall Meeting, San Antonio, Oct. 8-11, 1972.

18. Templeton, C. C., Richardson, E. A., Karnes, G. T., and Lybarger, J. H.: "Self-Generating Mud Acid," *J. Pet. Tech.* (Oct. 1975) 1199-1203.

19. Templeton, C. C., Street, E. H., Jr., and Richardson, E. A.: "Dissolving Siliceous Materials With Self-Acidifying Liquid," U.S. Patent No. 3,828,854 (1974).

20. Lybarger, J. H., Scheuerman, R. F., and Karnes, G. T.: "Buffer-Regulated Treating Fluid Positioning Process," U.S. Patent No. 3,868,996 (1975).

Chapter 10

Matrix Acidizing of Carbonates

10.1 Introduction

As indicated in Chapter 1, in the earliest acid treatments hydrochloric acid was dumped into the wellbore and pressure was applied to force the acid into the formation. Because of equipment limitations, early treatments were conducted at pressures below the formation fracture pressure. As high-pressure pumping equipment became available, treatment pressure was increased until most carbonate acid treatments today are conducted above the formation fracture pressure. Nonetheless, some reservoirs remain where acid treatments must be performed below the formation fracture pressure to be effective.

This chapter describes a matrix acid treatment of a carbonate and introduces a procedure for treatment design. Chemical reaction considerations are discussed in Chapters 3 and 4 and the fundamentals of matrix acidizing are presented in Chapter 8. The reader should consult these chapters for important background information.

10.2 Description of a Matrix Acid Treatment in a Carbonate

In a carbonate matrix acid treatment, the acid used (usually hydrochloric acid) is injected at a pressure (and rate) low enough to prevent formation fracturing. The goal of the treatment is to achieve more-or-less radial acid penetration into the formation to increase the apparent formation permeability near the wellbore. A more detailed description of the treatment and procedures for predicting the stimulation ratio are given in Chapter 2.

The treatment usually involves acid injection followed by a sufficient afterflush of water or hydrocarbon to clear all acid from well tubular goods. A corrosion inhibitor is added to the acid to protect wellbore tubulars. Other additives, such as antisludge agents, iron chelating agents, de-emulsifiers, and mutual solvents, are added as required for a specific formation (see Chapter 11).

10.3 Mechanism of Acid Attack

When acid is pumped into a carbonate (limestone or dolomite) at pressures below the fracture pressure, the acid flows preferentially into the highest permeability regions (that is, largest pores, vugs, or natural fractures). Acid re-action in the high-permeability region causes the formation of large, highly conductive flow channels called wormholes.[1-4] As discussed in Chapter 8, the creation of wormholes is related to the rate of chemical reaction of the acid with the rock. High reaction rates, as observed between all concentrations of HCl and carbonates, tend to favor wormhole formation.

Formation of a wormhole is illustrated in Fig. 10.1 for a linear core and Fig. 10.2 for a radial system. The photographs in Fig. 10.1 show the end of a limestone core into which 1-percent HCl was injected. On initial acid contact, several large pores formed; in time, the number of pores being enlarged decreased until only a few were accepting acid. In this example, the large pore shown in the lower left corner of the photographs ultimately accepted nearly all the acid and formed a wormhole extending through the total length of the core.

Acids normally used in field treatments are highly reactive at reservoir conditions and tend to form a limited number of wormholes. This conclusion is based on laboratory experiments and theories of matrix acidizing (see Chapter 8). Neither theoretical nor experimental studies can predict the number, size, or length of wormholes, although wormhole formation can be demonstrated. If the acid reaction rate is very high, the theories predict that only a few wormholes will form. A low reaction rate favors the formation of several small-diameter wormholes. Experiments conducted by Imperial Oil Ltd. in Canada[4] appear to verify this prediction. In these experiments, acid was injected into a cylindrical model prepared from carbonate rock. X-ray photographs then were taken to determine changes caused by the acid. All tests were performed at room temperature and with atmospheric pressure at the core outlet. Typical results from these experiments are illustrated in Fig. 10.2. The upper photograph (characteristic of a fast reaction rate) shows development of one wormhole. The lower photograph, in which the limestone was precoated with a surface-active retarding agent before acid injection, shows the development of multiple wormholes with subsequent branching. The single wormhole case is more representative of commonly encountered treating conditions.

TABLE 10.1—MAXIMUM ACID PENETRATION DISTANCE — NO
FLUID LOSS
(0.25 bbl/min, 40 perforations)

Number of Wormholes/ Perforation	Maximum Acid Penetration (ft)
1	225.
2	112.
3	75.
4	56.
5	43.
10	23.

TABLE 10.2—MAXIMUM WORMHOLE LENGTH — FLUID LOSS
LIMIT
(0.25 bbl/min, 40 perforations, 1 wormhole/perforation)

Formation Permeability (md)	Maximum Wormhole Length, L (ft)
1	8.9
5	2.0
25	0.5
100	0.1

10.4 Prediction of Radius of Reaction

Wormhole length normally is controlled by the rate of fluid loss from the wormhole to the formation matrix. Studies described by Nierode and Williams[5] show that the maximum wormhole length ranges from a few inches to a few feet. Wormhole length can be substantially increased by reducing the rate of fluid loss from the wormhole to the formation (in theory, a wormhole length of 10 to 100 ft is possible). The range of theoretically possible wormhole lengths calculated, assuming no fluid loss from the worm-hole and assuming fluid losses are controlled by the spent acid viscosity, are presented in Tables 10.1 and 10.2. Details of the calculations used to derive these data are given in Ref. 5.

The rate of fluid loss from a wormhole often can be reduced with a fluid-loss additive, thereby increasing wormhole length. The type of additive and concentration used must be selected carefully. Too much additive can plug the formation and prevent completion of the treatment. Too little additive will not be effective. One way of evaluating additives is to use the fluid-loss test described in Chapter 5.

a. 80 seconds exposure

b. 400 seconds exposure

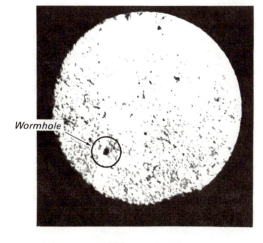

c. 1000 seconds exposure

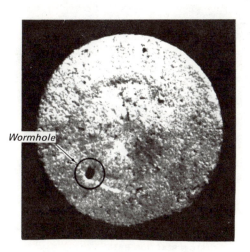

d. 1900 seconds exposure

Fig. 10.1—Development of a wormhole (1-percent HCl-Indiana limestone).

Normally, the most effective additives are solids or acid-swellable polymers used as acid-fracturing, fluid-loss additives. However, emulsified acids, because of their high viscosity, often will give better results than hydrochloric acid alone. Chemically retarded acids normally are not better than hydrochloric acid alone in a matrix treatment because they do not control the rate of fluid loss from a wormhole.

Laboratory data relating wormhole growth rate (in ft/min) to the concentration of a commercial oil-soluble additive were reported by Nierode and Kruk.[6] Data in Table 10.3 illustrate the change in growth rate with additive concentration. At 200°F and a differential pressure of 500 psi (across a 12-in. core), the maximum growth rate was observed at a concentration of 15 lb of additive/1,000 gal of acid. At a concentration of 50 lb/1,000 gal, the additive did not accelerate the wormhole growth rate, and at 100 lb/1,000 gal actually restricted the growth rate.

As discussed in Chapter 2, the productivity increase that can result from a matrix acid treatment in a carbonate normally is limited to damage removal. Without a fluid-loss additive, acid penetration distances will be limited to a few feet at most. The maximum stimulation expected from a matrix treatment will be about 1.5-fold above damage removal (Fig. 2.3). With an effective fluid-loss additive, the maximum wormhole length could exceed 10 ft (Table 10.1) and stimulation ratios slightly above twofold are possible, although seldom realized.

The exact stimulation ratio from matrix acidizing of a carbonate cannot be predicted because the number and location of the wormholes cannot be predicted. For these reasons, predictions using procedures described in Chapter 2 should be considered the maximum attainable.

10.5 Acids Used in Matrix Treatments

Because wormhole length normally is limited by fluid loss and not by acid reaction rate, organic acids, hydrochloric acid, acid mixtures, and chemically retarded acids should give comparable wormhole lengths and stimulation ratios when injected at matrix rates. A viscous, emulsified acid or hydrochloric acid containing an effective fluid-loss additive is preferred because both allow some control over the rate of fluid loss to the formation. In low-permeability formations, however, it often is not feasible to use these acids because of their low injectivity. In these formations, 28-percent HCl normally is preferred. If hydrochloric acid cannot be effectively inhibited to limit corrosion at formation temperature, formic acid is preferred, although acetic acid could be used (see Chapter 3 for an additional discussion of acid type and areas for application).

10.6 Design Procedure for Matrix Acidization of Carbonates

The design of a matrix acid treatment for a carbonate formation consists of specifying acid type and volume, and the maximum injection rate and pressure that can be used without fracturing the formation. A design procedure and an example design are given below (note that the first three

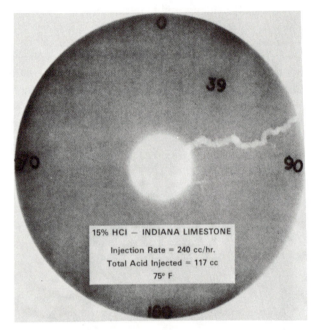

FAST REACTION RATE

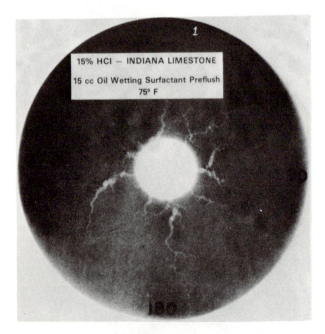

SLOW REACTION RATE

Fig. 10.2—Typical wormhole patterns formed in matrix acid treatments in carbonates.[4]

TABLE 10.3—EFFECT OF FLUID-LOSS ADDITIVE
CONCENTRATION ON WORMHOLE GROWTH RATE

Additive Concentration* (lb/1,000 gal)	Wormhole Growth Rate (ft/min)
0	0.021
15	0.052
50	0.026
100	0.008
200	0.002

*A commercial oil-soluble acid, fluid-loss additive reported to be a mixture of oil-soluble resins.[5]

steps in this procedure are the same as those used in the design of a matrix acid treatment for a sandstone).

Step 1. Determine the fracture gradient from prior fracturing treatments or with the approximate relation given as Eq. 7.2. (See Chapter 7, Section 7.2 for a discussion of this equation and its use.)

$$g_f \simeq \alpha + (\text{overburden gradient} - \alpha) \times$$

$$\frac{\text{reservoir pressure}}{\text{depth}} . \quad\ldots\ldots\ldots\ldots\ldots (7.2)$$

Step 2. Predict the maximum possible injection rate without fracturing using Eq. 9.1.

$$i_{max} =$$

$$\frac{4.917 \times 10^{-6} k_{av} h_n \, (g_f \times \text{depth} - \text{reservoir pressure})}{\mu \ln (r_e/r_w)} .$$

$$\ldots\ldots\ldots\ldots\ldots\ldots\ldots\ldots\ldots\ldots\ldots (9.1)$$

The injection rate to avoid fracturing must clearly be lower than i_{max}. As a practical matter, a rate 10 percent lower is suggested. Note that k_{av} is the average formation permeability, considering the effect of the damage zone, and not the undamaged formation permeability. In Eq. 9.1, permeability is expressed in millidarcies, pressure in psi, viscosity in centipoise, and thickness in feet giving i_{max} in barrels per minute.

The acid viscosity used in Eq. 9.1 can be estimated from Fig. 9.12.

Step 3. Predict the maximum surface pressure for which fluids can be injected without fracturing the formation.

$$p_{max} = (g_f - \text{acid hydrostatic gradient}) \, \text{depth}. \quad\ldots (9.2)$$

Data for the acid hydrostatic gradient can be obtained from Fig. 9.13.

During the treatment, surface pressure should be limited to pressures below this maximum, if possible. *If injection is not possible* without exceeding this pressure, it may be exceeded initially, but as soon as injection is started the rate should be reduced to maintain a surface pressure lower than p_{max}.

Step 4. Determine the volume and type of acid required. Inject from 50 to 200 gal of either 15- or 28-percent HCl per foot of interval perforated. *Exact acid volume and strength requirements cannot be predicted because of uncertainties in near-wellbore conditions and will vary from formation to formation*. In general, a larger volume of acid should be used in high-temperature wells or wells where deep damage is expected. When possible, use 28-percent hydrochloric acid and an effective acid fluid-loss additive in zones where acid can be injected at a realistic rate (at least 0.25 to 0.5 bbl/min). In high-permeability or naturally fractured formations, an emulsified acid often will give best results.

Example Design Calculation

To illustrate the use of the design procedure, design a treatment for an example well with the following characteristics: (1) $kh = 200$ md-ft, (2) perforated interval = 20 ft, (3) formation depth = 7,500 ft, (4) fracture gradient = 0.7 psi/ft (at initial pressure of 3,075 psi), (5) overburden gra-

dient = 1.0 psi/ft, (6) reservoir pressure = 2,000 psi, (7) acid viscosity at reservoir temperature = 0.4 cp, (8) drainage radius = 660 ft, and (9) wellbore radius = 0.25 ft.

Step 1. The fracture gradient is not known for the current reservoir pressure of 2,000 psi, but is known for the initial pressure of 3,075 psi. To estimate the current fracture gradient, compute the constant, α, using Eq. 7.2, the overburden gradient, and the fracture gradient data at *original* reservoir pressure.

$$0.7 = \alpha + (1 - \alpha) \times \frac{3,075}{7,500} .$$

Solving for α, we find

$$\alpha = \frac{0.29}{0.59} = 0.49,$$

and the *approximate* fracture gradient at *current* reservoir pressure is

$$g_f = 0.49 + (1 - 0.49) \times \frac{2,000 \text{ psi}}{7,500 \text{ ft}},$$

$$g_f = 0.49 + 0.14 = 0.63 \text{ psi/ft.}$$

Step 2. The maximum injection rate can now be estimated from Eq. 9.1 as

$$i_{max} = 4.917 \times 10^{-6} \, (200 \text{ md-ft}) \times$$

$$\left[\frac{(7,500 \text{ ft} \times 0.63 \text{ psi/ft} - 2,000 \text{ psi})}{0.4 \text{ cp} \ln (660/0.25)} \right]$$

$$i_{max} = 0.85 \text{ bbl/min (in practice, a maximum rate of}$$
$$0.77 \text{ bbl/min will be specified:}$$
$$0.9 \times 0.85 \text{ bbl/min} = 0.77 \text{ bbl/min)}$$

Step 3. The maximum surface injection pressure that can be used without fracturing the formation is

$$p_{max} = (0.63 - 0.43 \text{ psi/ft})(7,500 \text{ ft}),$$

$$p_{max} = 1,500 \text{ psi.}$$

Step 4. In this instance, an acid volume of 100 gal/ft will be assumed adequate to overcome damage. The job specification is as follows.

● Inject 2,000 gal of 28-percent HCl at an injection rate *not exceeding 0.77 bbl/min*. The surface pressure *should not exceed 1,500 psi*. When, or if, 1,500 psi is reached, the injection rate should be reduced. Corrosion inhibitors and other additives should be selected to protect for problems specific to the formation of interest. (See Chapter 11 for a discussion of additives.)

● Flush the treatment with more than one tubing volume of water. Put the well on production as soon as possible.

10.7 Novel Matrix Acid Treatments for Carbonates

Because hydrochloric acid reacts rapidly when contacted with either calcite or dolomite, a number of patents have been issued disclosing methods intended to reduce the reaction rate, thereby allowing acid to penetrate farther into the formation or to react more uniformly around the wellbore. The proposed techniques can be divided into two general classifications.

1. One classification includes techniques that involve the injection of chemicals (not acids) that react slowly, but at a

controlled rate to produce an acid. The acid product then may react rapidly with the reservoir minerals. These processes limit the reaction rate by producing the acid in situ at a controlled rate. They may have the further advantage of circumventing difficult corrosion problems because the chemicals from which the acid is derived may be less corrosive than the acid.

2. Another classification involves methods for retarding the hydrochloric acid reaction rate.

In-Situ Acid Formation

Hydrochloric Acid. Halogen derivatives of hydrocarbons may in some circumstances slowly hydrolyze in the presence of water to yield as products an alcohol and an inorganic acid. Many of these compounds are liquid at room temperature and, therefore, one might imagine a process using this hydrolysis reaction whereby HCl can be produced at a controlled rate. There is, however, one difficulty — namely, the solubility of halogen compounds in water is generally quite small, too small to produce in-situ an amount of acid capable of dissolving enough rock to increase productivity.

Dilgren[7,8] has proposed that a mutual solvent be used to produce a miscible solution of the halogen and water. One example is a solution of allyl chloride, water, and isopropyl alcohol that forms a single phase, provided a sufficient quantity of isopropyl alcohol is used. The reaction is

$$C_3H_5Cl + H_2O \rightleftarrows C_3H_5OH + HCl. \quad \ldots\ldots\ldots\ldots (10.1)$$

allyl chloride   allyl alcohol

If isopropyl alcohol is used in the proportion of nine volumes for each volume of water, then the reaction shown by Eq. 10.1 is first order in allyl chloride concentration and more than 4,000 minutes are required for one-half of the initial concentration of allyl chloride to react at a temperature of 200°F. This reaction time can be varied widely by changing the organic portion of the halogen (normal propyl chloride and tertiary butyl chloride are also possible halogens according to Dilgren[7]), the alcohol-to-water ratio in the solvent system, the nature of the alcohol used in the solvent system, and the temperature. Dilgren suggests that water be present in stoichiometric excess to assure that no unreacted halogens are solubilized in the oil because these particular compounds poison platinum-reforming catalysts used in oil refining.[7] It is important to recognize also that this system has the potential for serious environmental and health problems if not properly utilized.

Dilgren and Neuman[9] modified this process to eliminate the expensive mutual solvent. They proposed that the water and halogenated hydrocarbon be blended with an appropriate surfactant to form an emulsion. This water-external emulsion then is injected into the formation. The water and the hydrocarbon phases are intimately mixed within the pore structure and react at a controlled rate to produce HCl.

Organic acids. Perrine[10] has patented a concept in which a selected organic compound capable of producing an organic acid upon contacting water is dissolved in a liquid hydrocarbon. The hydrocarbon solution is injected into a carbonate formation, whereupon the organic compound diffuses from the hydrocarbon phase and reacts with connate water, producing an organic acid capable of dissolving the carbonate. One proposed system is a liquid hydrocarbon obtained by extraction of a petroleum fraction with a polar solvent (such as sulfur dioxide) mixed with acetic anhydride. On contacting connate water, the acetic anhydride is extracted to the water and then hydrolyzes to form acetic acid. The reaction of this acid with various carbonates was discussed in Chapter 3.

This technique was considered by Perrine to have several additional advantages over conventional treatments. Acid could be made to penetrate farther into the formation before reacting and the corrosion rate with well tubing would be reduced considerably.

Retarded Acids

Organic acids have been proposed for use by a number of investigators. Fatt and Chittum[11] developed a process whereby organic acids are mixed with a hydrocarbon to form a single-phase solution. When this solution was injected into the formation, it was envisioned that the acid would diffuse from the hydrocarbon phase into the connate water and subsequently react with the carbonate formation. Bombardieri and Martin[12] noted that certain mixtures of organic acids appear to react more slowly than HCl alone. They stated that 35-percent acetic acid will react more slowly than 10-percent acetic acid.

Gidley[13] has claimed the use of a concentrated organic acid (propionic), which when injected into the formation reacts to form calcium propionate that is insoluble in the concentrated organic acid. The calcium propionate subsequently is dissolved with a salt-water afterflush to provide the permeability increase. This system has been used successfully in high-temperature formations for damage removal. It also has been used (though not commonly) without an acid corrosion inhibitor.

Gidley and Tomer[14] extended this concept with a second patent for acid fracturing, using organic acid anhydrides.

Limitation of Novel Treatments

Some of the treatments discussed may have only marginal benefits over a conventional HCl matrix treatment because a matrix treatment normally is designed to overcome formation damage, as discussed in Chapter 2. HCl alone may accomplish this. Improved results could be expected for a novel treatment if it allowed a significant permeability increase for 10 to 20 ft, or more, around the wellbore (see Fig. 2.3).

References

1. Schechter, R. S. and Gidley, J. L.: "The Change in Pore Size Distribution From Surface Reactions in Porous Media," *AIChE Jour*. (May 1969) **15**, 339-350.

2. Rowan, G.: "Theory of Acid Treatments of Limestone Formations," *I. Inst. Petrol*. (1957) **45**, No. 431.

3. Guin, J. A.: "Chemically Induced Changes in Porous Media," PhD dissertation, U. of Texas (Jan. 1970).

4. Jazraui, Waleed: Private communication, Imperial Oil Ltd., Calgary.

5. Nierode, D. E.. and Williams, B. B.: "Characteristics of Acid Reaction in Limestone Formations," *Soc. Pet. Eng. J.* (Dec. 1971) 406-418; *Trans.*, AIME, **251**.

6. Nierode, D. E. and Kruk, K. F.: "An Evaluation of Acid Fluid Loss Additives, Retarded Acids, and Acidized Fracture Conductivity," paper SPE 4549 presented at the SPE-AIME 48th Fall Meeting, Las Vegas, Sept. 30-Oct. 3, 1973.

7. Dilgren, R. E.: "Acidizing Oil Formations," U.S. Patent No. 3,215,199 (Nov. 2, 1965).

8. Dilgren, R. E.: "Acidizing Oil Formations," U.S. Patent No. 3,297,090 (Jan. 10, 1967).

9. Dilgren, R. E. and Newman, F. M.: "Acidizing Oil Forma-tions," U.S. Patent No. 3,307,630 (March 7, 1967).

10. Perrine, R. L.: "Method of Acidizing Petroliferous Forma-tions," U.S. Patent No. 2,863,832 (Dec. 9, 1958).

11. Fatt, I. and Chittum, J. F.: "Method of Treating Wells with Acid," U.S. Patent No. 2,910,436 (Oct. 27, 1959).

12. Bombardieri, C. C. and Martin, T. H.: "Acid Treating Pro-cess," U.S. Patent No. 3,251,415 (May 17, 1966).

13. Gidley, J. L.: "Method for Acid Treating Carbonate Forma-tions," U.S. Patent No. 3,441,085 (April 29, 1969).

14. Gidley, J. L. and Tomer, F. S.: "Acid Fracturing Process," U.S. Patent No. 3,451,818 (July 1, 1969).

Chapter 11

Acid Additives

11.1 Introduction

All commercial acids used in well stimulation require a corrosion inhibitor to reduce the rate of acid attack on the wellbore tubular goods. Other materials sometimes are added to acid (1) to eliminate emulsion formation, (2) to alter formation wettability for improved acid attack or better cleanup, (3) to reduce friction drop through tubular goods allowing higher pumping rates, (4) to reduce the rate of fluid loss from a fracture, (5) to divert acid flow from one zone to another allowing more uniform treatment, (6) to complex iron and prevent its precipitation, and (7) to avoid sludging in certain highly asphaltic oils.* This chapter describes briefly the function of each additive type and introduces some materials used. Many of these materials are proprietary and definitive chemical descriptions often are unavailable. Testing procedures to be followed in evaluating the additives are detailed in API RP 42 and therefore will be only mentioned in this chapter. All tests should be conducted with the final acid and additive formulation and also with reacted acid.

11.2 Corrosion Inhibitors

Foremost among acid additives are corrosion inhibitors. Not only is the cost of the corrosion inhibitor often a significant portion of the total acidizing cost, especially if high bottom-hole temperatures are encountered or long acid-pipe contact times are anticipated, but in some cases, the choice of the acid treatment itself is governed by the selection of the most economical means for controlling corrosion, without the corrosion inhibitors impairing after treatment productivity. An explanation of the mechanism of acid corrosion inhibition and techniques for evaluating acid-corrosion inhibitor performance are described briefly below. Factors important in inhibitor evaluation and practical suggestions to help select corrosion control inhibitors then are presented.

Mechanism of Acid Corrosion Inhibition

In acid corrosion, electrolytic cells are thought to be set up on the metal surface. As illustrated in Fig. 11.1, metallic iron goes into solution at anodic sites and electrons are liberated at cathodic sites, reducing hydrogen ions to gaseous hydrogen. To be effective, a corrosion inhibitor must reduce the reaction rate at the anodic or cathodic site, or at both sites.

Broadly, two classes of inhibitors (distinguished by the manner in which they inhibit corrosion) are found to be useful. One class (the anodic type) functions by sharing electrons from the inhibitor molecule with anodic sites on the metal surface. The bond thus established terminates the reaction at that site. The second class (the cathodic type) forms a protective film by attachment of the cationic inhibitor to the cathodic area of the metal surface through electrostatic attraction.

Inhibitor effectiveness is a function of its ability to form and maintain a film on the steel surface. Therefore, factors that reduce the number of inhibitor molecules adsorbed will reduce inhibitor effectiveness. Perhaps the most important limiting factor is temperature. At a high temperature, the acid corrosion rate increases and the ability of the inhibitor to adsorb on the steel surface decreases. For these reasons, inhibitors for high-strength acid at temperatures above about 250°F are expensive and difficult to formulate.

Many compounds have been evaluated experimentally for their ability to reduce acid corrosion. Commercially available inhibitors are usually complex mixtures of organic compounds. The patent literature is extensive on this subject and the whole area is highly proprietary. Some more common chemicals included in inhibitors are the thiophenols, nitrogen heterocyclics, substituted thioureas, rosin amine derivatives, acetylenic alcohols, and arsenic compounds.[1]

For our purpose, neither the chemical nor the electrolytic aspects of corrosion inhibition are of great concern, so long as the materials are effective for reducing acid corrosion. Unfortunately, effectiveness must be determined by laboratory tests, which are not yet standardized throughout the industry. The user must, therefore, be aware of the factors in inhibitor testing that significantly influence test results to assure he obtains adequate protection at lowest cost.

---

*Caution: Acids and their additives are often hazardous to those handling them. Service companies have toxicity data on all products. This information should be consulted to assure adequate precautions are taken in the field.

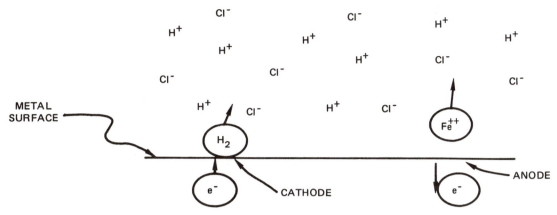

Fig. 11.1—Electrolytic cells are set upon metal surface by acid corrosion.

Evaluation of Acid Corrosion Inhibitors

The laboratory evaluation of a corrosion inhibitor generally involves subjecting a coupon of the metal to be protected to the acid to be used.[26] A metal coupon normally is inserted in a heated pressure vessel (autoclave) containing the acid and corrosion inhibitor being evaluated. The amount of corrosion with the coupon is determined by weighing it before and after the test, or periodically during the test. Inhibitor effectiveness is expressed in terms of metal loss per unit area of metal exposed per unit time (often lb/sq ft/D). Control tests are used to determine weight loss in the acid without inhibition. In addition to determining the rate of weight loss, the tendency of the acid to form pits, rather than reacting uniformly with the surface, is usually noted.

Many factors influence the corrosion rate measured in such tests.[27] Of major importance are (1) the amount of agitation, (2) metal type, (3) exposure time, (4) temperature, (5) acid type and concentration, (6) inhibitor type and concentration, (7) metal area/acid volume ratio, (8) pressure, and (9) the presence of other additives, such as surfactants or mutual solvents.

An increase in *agitation* increases the corrosion rate, as demonstrated in Fig. 11.2[2] Hence, static tests (i.e., no agitation) make the performance of the inhibitor appear more favorable than it is likely to be when acid is pumped through tubular goods.

The effect of *metal type* is significant. In Fig. 11.3, results are shown for five commercial corrosion inhibitors to illustrate the difference in response obtained, depending on whether they are inhibiting corrosion on N-80, J-55, or P-105 tubular goods[2] (inhibitors were not identified). The API designations for steels are inadequate to distinguish differences in corrosion rates as Smith[3] has pointed out, because they identify yield strength, whereas, corrosion characteristics are determined primarily by chemical composition. Because yield strength specifications may be achieved by either physical or chemical modifications in manufacturing, it is essential that the coupons tested be from

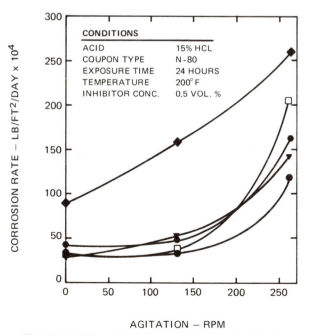

Fig. 11.2—Effect of agitation on acid corrosion inhibitor effectiveness.[2]

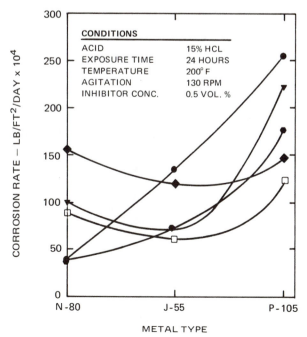

Fig. 11.3—Effect of metal type on acid corrosion rate.[2]

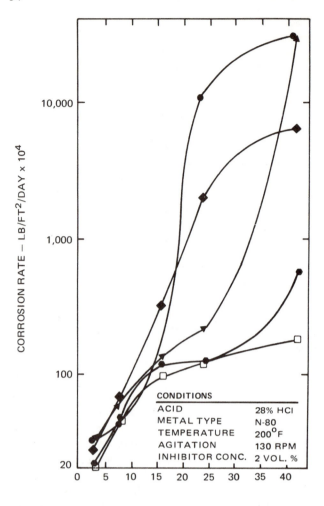

Fig. 11.4—Effect of exposure time on average acid corrosion rate.[2]

CONDITIONS

ACID	28% HCl
METAL TYPE	N-80
TEMPERATURE	200°F
AGITATION	130 RPM
INHIBITOR CONC.	2 VOL. %

a representative sample of the tubular goods to be protected.

Corrosion increases with *exposure time* are illustrated in Fig. 11.4 for the same five inhibitors. Inhibitors that offer similar protection for short exposure periods may provide widely different protection for longer periods.

Increasing *temperature* increases the acid corrosion rate (Fig. 11.5). The difficulty of corrosion control in acids at temperatures above 200°F is readily seen to be substantial.

Acid type and concentration also greatly influence the effectiveness of acid corrosion inhibitors (Fig. 11.6). Although corrosion in mud acid (12-percent HCl and 3-percent HF) at a given temperature is no more difficult to control than that in 15-percent HCl, corrosion with 28-percent HCl is much more difficult to control. The inhibitor concentration required for effective corrosion control depends on both temperature and type of metal goods as illustrated previously. Also, both the duration of protection and the degree of protection required (that is, the weight loss per unit area of metal goods) influence the concentration of inhibitor required. Curves illustrating this relationship generally are available from service companies for their own corrosion inhibitors. As noted previously, results from static tests will often underestimate the amount of inhibitors required, and must be used with care.

A more subtle variable is the *acid-volume-to-coupon-area ratio*. The use of a large acid-volume-to-coupon-area ratio appears to reduce the corrosion rate (Fig. 11.7). In effect, the greater acid volume brings more inhibitor for

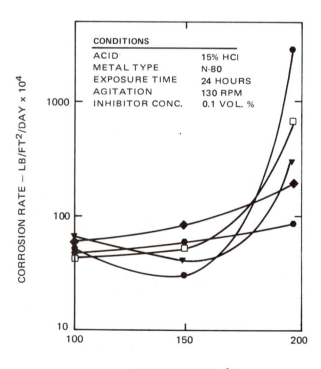

CONDITIONS

ACID	15% HCl
METAL TYPE	N-80
EXPOSURE TIME	24 HOURS
AGITATION	130 RPM
INHIBITOR CONC.	0.1 VOL. %

Fig. 11.5—Effect of temperature on corrosion inhibitor effectiveness.[2]

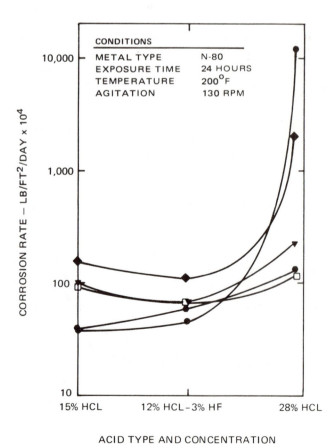

CONDITIONS

METAL TYPE	N-80
EXPOSURE TIME	24 HOURS
TEMPERATURE	200°F
AGITATION	130 RPM

Fig. 11.6—Effect of acid type and concentration on corrosion inhibitor effectiveness.[2]

adsorption on the coupon surface. Smith,[3] who has examined this matter critically, indicates that volumes of acid in excess of 75 ml/sq in. of surface are excessive compared with the common range of conditions found down hole.

The effect of *pressure* on corrosion inhibitor performance depends greatly on the inhibitor. Some inhibitors are most effective at high pressures; others are more effective at low pressures. A minimum test pressure of 1,000 psi appears adequate to eliminate from consideration inhibitors that are effective only at very low pressures.

The *additives* present in an acid system can modify the effectiveness of the corrosion inhibitor. Materials such as mutual solvents[4] or surfactants often alter the effectiveness of corrosion inhibition by preventing or aiding inhibitor adsorption. Additives, in general, should be included in the inhibitor test to determine their effect on inhibition and to assure that useful results are obtained.

Although weight loss corrosion is generally the prime consideration in acid corrosion inhibitor testing, the *pitting tendency* of the system should be determined. Smith[3] has suggested a useful scale for quantifying pitting tendency. Either this scale or some equivalent procedure (such as examination of the test coupons themselves) should be used to verify the absence of substantial pitting when an inhibitor is used.

Suggestions for Inhibitor Selection

The selection of inhibitor type and concentration can be done only after specification of the following treating and well conditions: (1) type and concentration of acid, (2) type of tubular goods to be exposed, (3) maximum pipe temperature, and (4) duration of acid-pipe contact. Generally, with these factors specified, service company information can be used to determine the inhibitor requirements needed to provide the required level of protection. Occasionally, other factors such as sulfide stress cracking must be considered.[28]

An important consideration is what is a practical level of protection. Most service company information is based on the assumption that a metal loss of 0.02 lb/sq ft of area can be tolerated during a treatment if no pitting occurs. Sometimes a figure as high as 0.05 lb/sq ft is assumed allowable. If these metal losses cannot be sustained without an adverse effect on well equipment, either a higher inhibitor concentration or a more effective inhibitor must be used.

Inhibition of acid corrosion at temperatures above 250°F for 8 hours or longer is one of the more difficult problems when acidizing deep wells.

In the past, arsenic inhibitors have been used for high temperature applications because of their effectiveness and low cost. Arsenic inhibitors, however, have the following disadvantages: (1) arsenic poisons catalysts used in oil refining, (2) arsenic compounds are not effective in the presence of hydrogen sulfide, (3) arsenic compounds are relatively ineffective in HCl concentrations above 17 percent, and (4) arsenic is a poison. Disposal of fluids from wells treated with an arsenic inhibitor in an environmentally acceptable manner is difficult at best.

Fortunately, continuing research has produced organic inhibitors that, while not as effective as arsenic, are useful at temperatures above 250°F. Sometimes their use requires a so-called inhibitor extender, generally an inorganic salt such as potassium iodide that synergistically improves inhibition. These combined organic-inorganic inhibitor systems, while expensive, can often be used at temperatures up to 400°F.

In some instances, inhibitor requirements can be reduced by careful treatment design. In acid fracturing treatments, for example, the corrosion rate may be reduced by precooling the tubing by injecting a water pad preceding the acid. In this way, temperatures may be reduced to the point that conventional (and less expensive) inhibitors offer sufficient corrosion control. This technique must be used with caution, however, since very high strength tubular goods (usually API Grade P-105 and higher) are susceptible to hydrogen embrittlement. This susceptibility is greater at temperatures below about 200°F than at higher temperatures.

11.3 Surfactants

Surface active agents are used in acid treating to demulsify acid and oil, to reduce interfacial tension, to alter formation wettability, to speed cleanup, and to prevent sludge formation. Caution should be exercised when adding surfactants to be sure that they are compatible with the corrosion inhibitor and other additives.

A demulsifying agent often is used in carbonate limestone acidizing treatments to prevent emulsion formation between the acid and formation crude oil. Regular acid containing a

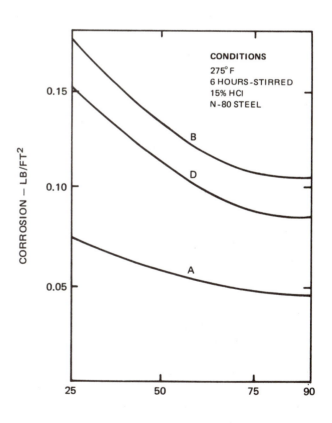

Fig. 11.7—Effect of acid volume/coupon area on corrosion rate in 15-percent HCl.[3] The results for three different inhibitors are shown.

demulsifying agent is sometimes designated *nonemulsifying (NE) acid*.[29] Organic amines and salts of quaternary amines (both cationic) are used in this application, as well as polyoxyethylated alkylphenols (nonionic). The tendency to form an emulsion is increased with small diameter fines (see Chapter 9, Section 9.5). The use of glycol ether mutual solvents helps prevent the formation of these solid stabilized emulsions, perhaps by preventing adsorption of the oil-wetting corrosion inhibitor on the fine solid particles.[5,6]

Surface active agents sometimes are added to acid to develop a so-called *low surface tension (LST) acid*,[29] although this application has diminished in recent years. Generally an alkyl aryl sulfonate (anionic) or ethoxylated alkylphenol (nonionic) surfactant at concentrations of 0.1 to 0.5 percent by volume is employed. The interfacial tension between acid or spent acid and the oil phase present is reduced to minimize capillary retention of fluids in the formation. Even though reduced interfacial tension can be achieved easily in the laboratory, Hall[5] has found that adsorption of the surfactant on formation materials near the wellbore limits the effectiveness of these surfactants in the field. He also determined that the presence of a mutual solvent often can extend the penetration depth of the surfactant into the formation.

Many of the special acids from service companies are proprietary mixtures of surfactants and hydrochloric acid. *Clean-out acids* are of this type.[29] They generally contain from ½- to 3-percent surfactant with from 5- to 15-percent hydrochloric acid. The surfactant used is often a mixture of an anionic (alkyl sulfonate) and nonionic (polyoxyethylated alcohol or ether) surfactants. Such acids are used widely in "hole-opener" treatments to initiate production from limestone formations that may not produce without acidization.

Antisludge agents appear to be required for effective acid treatment in fields containing heavy asphaltic oils, such as those found in parts of California, the Rocky Mountain states, and Canada. The sludges formed when these oils contact acid can plug the formation and restrict production after treatment. This problem can sometimes, but not always, be reduced by adding surface active chemicals such as alkylphenols, fatty acids, and certain oil-soluble surfactants to the acid. The need for, and the specific antisludge agent to be used, generally must be determined in laboratory tests with the crude oil and acid system to be employed.[7]

Certain surfactants are proposed as *acid reaction rate retarders* in acid fracturing treatments. For this application, surfactants are selected to adsorb on the rock surface, making it oil-wet and thereby presumably creating a physical barrier for acid transfer to the fracture surface. Chemical retardation by this mechanism often uses an oil preflush containing the additive, but the retarding agent may also be added with the acid. Published work indicates that several cationic, anionic, and nonionic surfactants have been evaluated for their rate-retarding abilities. Anionic surfactants, such as the alkyl phosphates, sulfonates, and taurates, also have been considered. Acid retardation by this mechanism (discussed in Chapter 5, Section 5.3) normally is not effective under fracturing conditions where high injection rates are employed.

11.4 Mutual Solvents

Mutual solvents were first mentioned in Chapter 9, which discussed their use in minimizing adsorption of oil-wetting surfactants on formation solids, thereby reducing the potential for productivity restrictions after acidization by relative permeability alteration or formation of solids stabilized emulsions. Mutual solvents also were discussed in the section on surfactants. We have tried to describe what is meant by a mutual solvent and to point out the usefulness these are thought to have in acid treatments.

Mutual solvents are materials that have appreciable solubility in both oil and water. Many chemicals, including alcohols, aldehydes, ketones, ethers, and others, have this property. In oilfield applications, the term "mutual solvent" is normally used to describe a glycol ether. The glycol ether most frequently used in sandstone acidizing is ethylene glycol monobutyl ether (EGMBE). EGMBE, in addition to its mutual solubility, reduces interfacial tension between oil and water, acts as a solvent for solubilizing oil in water, acts as a detergent capable of removing oil-wetting materials from surfaces that otherwise would be water-wet, and, finally, improves the action of surfactants and emulsifiers in contact with formation materials. The oil solubility of the glycol ethers is increased by increasing the length of the hydrocarbon end of the molecule. This change generally reduces its water solubility. Water solubility can be increased by reducing the length of the hydrocarbon chain or increasing the degree of ethylene oxide substitution.

Empirically, EGMBE has been found to have a suitable balance of water and oil solubility useful in reducing emulsification and expediting cleanup in sandstone acidizing treatments. Gidley[6] reported that the productivity of oil wells in sandstone formations treated with HF-HCl is increased by five- to sixfold over that found with regular HF-HCl treatments, if as much as 10-percent EGMBE is used in the diesel oil afterflush of the treatment. Taylor and

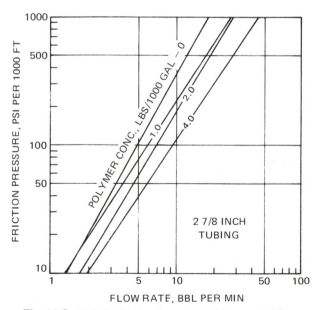

Fig. 11.8—Tubular friction pressure for 28-percent HCl containing a typical polyacrylamide friction reducer.

Plummer[8] have also demonstrated improved results on gas wells when a mutual solvent was used in the treatment.

Sutton and Lasater[9] have examined the role of mutual solvents in systems containing surfactants or corrosion inhibitors. They found that the mutual solvent reduces the adsorption of surface active materials on formation solids. Hall[5] has shown that the addition of a mutual solvent to a surfactant system increases the depth of penetration of the surfactant in laboratory sand packs. He evaluated many mutual solvents from the standpoint of their ability to reduce the adsorption of surface active materials on sand and concluded that EGMBE and a modified (proprietary) glycol ether were the most effective.

Although the greatest use of mutual solvents has been in sandstone acidizing, EGMBE has also been used in carbonate acidizing. Dunlap[10] described the process of preflushing an oil-containing limestone reservoir with EGMBE before injecting hydrochloric acid. The EGMBE acts as a cleaner and oil remover to improve the effectiveness of the acid treatment.

11.5 Friction Reducers

In an acid fracturing treatment, it is often desirable to pump the pad fluid and/or acid at the maximum rate allowable while maintaining the surface pressure below a fixed limit, or to minimize the horsepower required to pump at a specified rate. Chemicals that when dissolved in the fluid reduce the fluid's frictional pressure drop through well tubing are called friction reducers.

Friction reducers are normally organic polymers that convert the fluid from a Newtonian fluid (constant viscosity at all shear rates) to a non-Newtonian fluid (viscosity varies with shear rate). Typical behavior of the frictional pressure for acid after addition of a friction reducing additive is shown in Fig. 11.8 as a function of both the flow rate through 2⅞-in. tubing and polymer concentration. This figure shows that as the polymer content is increased, the friction pressure decreases from that for the base fluid. For example, at 10 bbl/min, the addition of 4 lb/1,000 gal of acid gives about a threefold reduction in pressure drop. Those polymer additives that reduce turbulent friction at relatively low concentrations often will yield pressure drops greater than for acid alone when used in higher concentration. For polymers now in use, the minimum pressure drop usually occurs at concentrations ranging from about 1 to 20 lb/1,000 gal of fluid.

As discussed in the next section, many polymers used as friction reducers also are often added to give the pad fluid, or acid, improved fluid-loss characteristics. To be effective as a fluid-loss additive in acid, the polymer content often will have to be from 50 to 200 lb/1,000 gal of acid. Generally, treatment results will be best when the fluid-additive system is specified to give adequate fluid-loss control, and treatment volume and injection rate then are varied to obtain the desired stimulation ratio.

Many different additives are used as friction reducers. Generic classifications of additives for water- and oil-based pad fluids and for acids are given in Table 11.1. A general description of fluids used as a pad before acid injection is

TABLE 11.1—FRICTION REDUCERS

Fluid Type	Generic Classification of Additive
Water-based pad	Guar Polyacrylamide Cellulose
Oil-based pad	Polyisobutylene Fatty acids Crosslinked organic polymers
Acid	Guar Gum karaya Polyacrylamide Cellulose

given in Chapter 7. Guar, karaya, celluloses, and polyacrylamides are polymers that can have different properties, depending on molecular weight, chemical composition, and so forth. Some polyacrylamides are extremely efficient friction reducers for acid and can greatly reduce the frictional pressure drop at concentrations of 1 to 4 lb/1,000 gal of acid (see Fig. 11.8). When friction pressure alone is important (fluid loss is not important), these additives will reduce the friction pressure at the lowest cost. The effectiveness of most polymers is limited because they degrade rapidly in acid, particularly at elevated temperatures (see Fig. 5.12, for example).

11.6 Acid Fluid-Loss Additives

To create a long, wide fracture, it is necessary to minimize the rate of fluid loss from the hydraulically induced fracture to the formation (see Chapter 5, Section 5.3). Fluid-loss additives, to be effective, often are composed of two agents: (1) an inert, solid particle that can enter the formation pores but will bridge near the fracture surface (Fig. 11.9a) and (2) a gelatenous material that will plug the pores in the solid granular material (Fig. 11.9b).

When a fluid-loss additive is included in the pad fluid or acid, the volume of fluid loss per unit of fracture surface area has the characteristic behavior illustrated in Fig. 11.10. Often the rate of fluid loss is modeled with a spurt volume, V_{spt}, and a fluid-loss coefficient, C_w, as discussed in Chapter 6, Section 6.3. Test procedures for evaluating fluid-loss

A. SOLID PARTICLES ENTER AND BRIDGE IN PORES

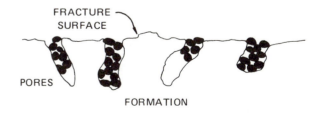

B. SOLID ADDITIVE GELATINOUS MATERIAL PLUGS PORES IN

Fig. 11.9—Behavior of typical fluid-loss additives.

additives used with pad fluids are discussed in detail by Howard and Fast.[11] When evaluating additives for use in acid while treating carbonate formations, fluid-loss test procedures should include the following restrictions: (1) a carbonate core must be used so acid reaction occurs with the core while the test is in progress, (2) a 6- to 18-in.-long core is preferred so that wormholes formed by acid reaction do not penetrate rapidly through the core, and (3) both the core and acid must be at the temperature expected in the fracture during the treatment. Erroneous results are often obtained if these guidelines are not followed.

Commonly used fluid-loss additives are listed in Table 11.2. In aqueous pad fluids, a combination of solid additive (most often silica flour) and a polymer (guar, cellulose, or polyacrylamide) is usually selected. When water without a polymer additive is to be used, an inert solid coated with a guar-type material is sometimes selected. Acid additives include acid-soluble organic polymers that swell and become soft when contacted by acid, thereby giving both solid and gelatenous characteristics. Other additives are similar to those chosen for use in water-based pad fluids.

There are several limitations associated with fluid-loss additives used in an acid treatment[14] that include the following.

1. *The fluid-loss additive deposited from the pad fluid is effective only for a short time once it is contacted by an acid that does not contain an effective fluid-loss additive.* This loss in effectiveness occurs because there is a continuing flow of acid through the filter cake deposited from the pad fluid. This acid eventually will destroy the rock matrix on which the additive from the pad is deposited, allowing the rate of fluid loss to increase (Fig. 11.11). If the acid contains an effective fluid-loss additive, this increase in rated acid fluid loss can be prevented.

2. *It is difficult to attain the same degree of fluid-loss control with an acid as with an inert fluid because acid reaction destroys the matrix on which the filter cake is being*

TABLE 11.2—COMMONLY USED FLUID-LOSS ADDITIVES

Fluid Type	Solid Additives	Gelatenous Additives
Aqueous pad	Silica flour Calcium carbonate Organic polymer Inert solid coated with guar-type material	Guar Cellulose Polyacrylamide
Hydrocarbon pad	Inert solid coated with organic sulfonate	
Acid	Acid-swellable solid Organic resin Silica flour Organic polymers	Guar Karaya Cellulose Polyacrylamide Polyvinyl alcohol

deposited. This means that much higher additive concentrations (50 to 200 lb/1,000 gal of acid) are required to achieve the same degree of fluid-loss control possible with a nonreactive fluid.

3. *Most fluid-loss additives cannot effectively seal formation fractures or vugs that may be intersected by the hydraulically induced fracture.* Since carbonate formations often contain these features, it is extremely difficult to be sure that fluid-loss characteristics measured on a relatively homogeneous core are representative of field behavior; indeed, they often are not. This means the induced fracture is often narrow and the acidized fracture length is shorter than predicted.

To minimize the effect of fluid loss to naturally occurring fractures, the following techniques occasionally have been used. (1) Small diameter silica sand (about 300 mesh) has been added to the pad in an attempt to bridge the natural fractures. (2) Treatments have been designed for alternately injecting pad fluid and acid, rather than injecting all the pad and then the acid. (3) Very viscous pad fluids have been used. The effectiveness of these various techniques has not been evaluated quantitatively; therefore, it is not possible to quantify any benefits that may occur.

4. *Acid reacts with all the polymeric additives used to control fluid losses, particularly at high temperatures.* This

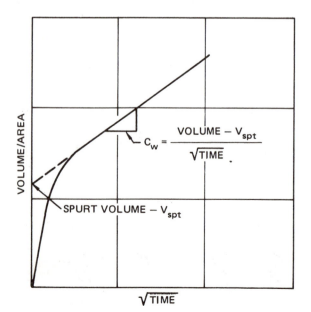

Fig. 11.10—Typical fluid-loss test results.

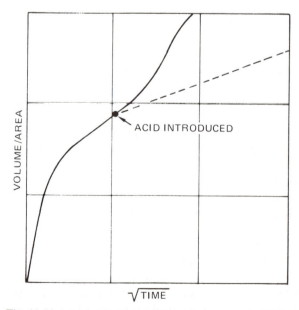

Fig. 11.11—Typical fluid-loss test results when acid without a fluid-loss additive follows an efficient pad fluid.

reaction destroys the polymer and renders it ineffective in the fluid-loss process. This is particularly true with guar, karaya, polyacrylamide, and cellulose polymers (incidentially, weak acids are often used to break these polymers in hydraulic fracturing applications). Polymers such as the polyvinyl alcohols, and perhaps crosslinked polyacrylamides, promise better high temperature stability in acid, although their use is not widespread.

As indicated by the four limitations discussed above, control of fluid losses with acid, particularly in naturally fractured formations, remains one of our most serious limitations. Research efforts to develop new additives should be encouraged. Any new materials should be thoroughly evaluated, using realistic test procedures before use.

11.7   Diverting Agents

To obtain maximum stimulation from a matrix or fracture acidizing treatment, it is necessary to treat the total productive interval. When several sands are open to the wellbore or the section to be treated is massive (illustrated in Chapter 7, Figs. 7.1 through 7.3), it is often necessary to divide the treatment into stages. A technique that forces each stage to go into a different zone is used during the procedure to assure treatment of the total productive interval. Flow often can be effectively diverted using down-hole equipment, such as packers. However, since using a workover rig to move packers can significantly increase job costs, techniques for separating liquid stages without packers have been developed. In these techniques, each fluid stage is followed by material that will temporarily plug the zone just treated, thereby diverting the next stage of acid to an untreated zone. Because the requirements for matrix and fracture-acidizing diverting agents differ, and because different diverting agents are used, they are discussed separately in the following sections. Mechanical separation techniques were discussed in detail by Howard and Fast in the monograph *Hydraulic Fracturing* (Section 7.5 and 8.8) and will not be reviewed here. A good historical review of the development and use of diverting techniques was presented by Harrison.[12]

Matrix Acidizing Diverting Agents

When used in matrix acidizing, diverting agents are designed to bridge at the formation pores and to function in much the same way as fracturing fluid-loss additives (Fig. 11.9). Chemicals used for matrix diversion include finely ground inert organic resins, solid organic acids, deformable solids, mixtures of waxes and oil-soluble polymers, acid-swellable polymers, and mixtures of inert solids (silica flour, calcium carbonate, rock salt, oil-soluble resins, etc.) with a water-soluble polymer (guar, gum, karaya, cellulose, polyacrylamide, and so forth).

These diverting agents often can be used to divert acid flow without damaging the formation, *if used sparingly*. Tests in which acid containing the additive is injected simultaneously into cores of different permeability (representing an open-hole completion) or into cores with a drilled hole (representing a perforation) generally show the behavior illustrated in Fig. 11.12. (1) The diverting agent first enters the zone of greatest flow capacity (the Berea core in this case) and causes a decrease in flow rate into the zone. (2) The flow rate into the lower permeability zone (the Bandera core in this case) is not appreciably altered until the rate of flow into each zone is essentially equal. (3) Once the flow rate into the zones is about equal, the injectivity into each decreases as additive injection continues.[14] The ideal quantity of diverting agent, therefore, appears to be that which just reduces injectivity into the high-permeability zone so that it is equal to the injectivity into the low permeability zone. If too much additive is used, injectivity into all zones can be drastically reduced.

Studies similar to the test just described lead to the following general recommendations when using matrix acidizing diverting agents.

1. Hard inert solids alone usually are not effective diverting agents — apparently because a material for plugging openings between the hard solid particles is not present.

2. Mixtures of hard and soft particles can be effective diverting agents. An example is the mixture of finely ground oil-soluble resins that contains particles of different sizes and hardness, mixtures of waxes, and a ground oil-soluble resin.[14, 16]

3. Acid-swellable polymers are effective diverting agents; however, only a small quantity of additive is required and care is necessary to prevent excessive use.[13, 14]

4. The matrix diverting agents do not completely divert flow from one zone to another; they tend to equalize the flow rate into each zone.

5. Agents soluble in acid are generally ineffective be-

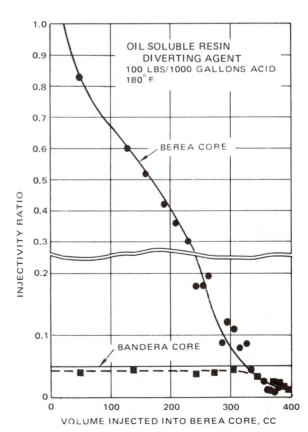

Fig. 11.12—Test of a typical oil soluble diverting agent.

cause they are quickly dissolved by acid in the stage following the diverting plug.[14]

6. The greatest problem when using diverting agents in a matrix acid treatment may be the use of too much additive. When excess additive is used, injectivity can be greatly reduced and removing the additive after the treatment may be difficult. Also, in an effort to maintain injectivity when too much additive is used, the treating pressure may increase to a level that fractures the formation, thereby reducing treatment effectiveness or causing high water or gas production ratios after the treatment.

Acid-Fracturing Diverting Agents

To be effective in an acid fracturing treatment, the diverting agent must either bridge at the perforation or immediately upon entering the fracture. The bridge formed in this fashion must have a low permeability and be strong enough to stand up under a differential pressure of several hundred psi. The agent must be removed easily from the fracture or perforation when the well is returned to production.[11, 17]

Ball sealers (developed as a diverting agent for acid and proppant fracturing treatments) are designed to seal at the entrance of a perforation, thereby stopping flow into the perforation.[18, 20] When used in "pseudo" limited entry or limited entry treatments (described in Chapter 7, Section 7.8), ball sealers are probably the most effective fracturing diverting agent.[20-22] However, they are not effective in open-hole completions and often cannot be used effectively in wells with large numbers of perforations. In these instances, granular diverting agents are often employed. Details on the use of ball sealers were given by Howard and

Fast[11] in the *Hydraulic Fracturing* monograph and therefore are not repeated here.

Chemicals used as fracturing diverting agents are similar in composition to those used in matrix diversion. The primary difference is that much larger particles are included in the mix. Commonly used additives include rock salt, solid organic acids, ground inert organic resins, oil-soluble waxes, and various combinations of these additives. In all cases, a broad particle-size distribution is required, including large particles that will bridge on the fracture, and smaller particles to fill the openings between the large particles. Best results often are obtained when the solid additive is placed in a viscous gelled oil or water-based fluid.

Studies of the various fracturing diverting agents by Gallus and Pye[16, 23] and, more recently, by Novotny[24] have shown that the oil-soluble, wax-polymer agents are among the most effective diverting agents available. Data are presented in Fig. 11.13 comparing performance of a wax-polymer blend, rock salt, and naphthalene in laboratory tests to illustrate this point. Comparative field results from the same study (Fig. 11.14) tend to verify the laboratory data. General conclusions from studies of the particulate additive are as follow.

1. Water- or acid-soluble diverting agents are often less effective than nonwater-soluble agents since acid from stages injected after the plug can dissolve the plugging materials. This effect is most pronounced at high bottom-hole temperatures with organic materials like naphthalene and benzoic acid.

2. Laboratory tests that do not include a study of how long a plug can hold while contacting acid are not representative of field conditions and often give overly optimistic results.

3. Water-soluble diverting agents can be made effective if the flow rate through the plug is low enough. Novotny found that a 50:50 mixture of rock salt and paraformaldehyde flakes placed in a saturated brine containing 60 lb of guar and 100 lb of silica flour/1,000 gal can be effective at formation temperatures as high as 300°F. Field results in water injection wells verified the laboratory data.[24]

4. The oil-soluble, wax-polymer agents are among the most effective diverting agents for use in oil-based fluids or oil-producing formations.[16, 23, 25] Care must be taken to select the wax blend for the appropriate formation temperature. Often, it is possible to select a wax that will be solid at bottom-hole treating temperatures, but will melt when the fracture returns to the native formation temperature, thereby easing cleanup (see Chapter 6, Section 6.4 for a discussion of wellbore and fracture temperature during a fracturing treatment).

11.8 Complexing Agents

When appreciable quantities of iron in the ferric ionization state (Fe^{+++}), as opposed to the more typical ferrous state (Fe^{++}), are dissolved by the acid, iron precipitation and permeability reductions can occur after acidization. Sources of iron include (1) corrosion products found on the walls of the tubular goods, (2) pipe scale, and (3) iron in mineralogical form in the formation. The ratio of ferric to ferrous iron

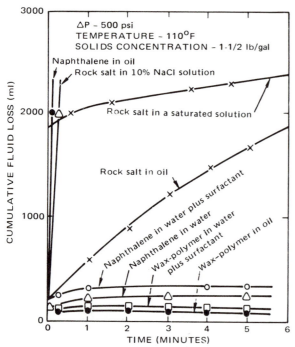

Fig. 11.13—Comparison of fracture sealing capacities of wax-polymer, naphthalene, and rock salt diverting agents in various carrying fluids.[16]

governs precipitation. Ferric iron precipitates at a pH of about 2.2, while ferrous iron precipitates at a pH of about 7. Since spent acid solutions seldom rise to a pH above 6, precipitation of ferrous iron is seldom a problem. Precipitation of ferric iron occurs as acid is spent and the pH rises above 2.2[25]

Precipitation of ferric iron during acidizing is by no means a common problem. Ferric iron, in the presence of iron (steel) pipe and hydrochloric acid, rapidly converts to ferrous iron; therefore, the normal dissolution of pipe rust and scales does not necessarily lead to iron precipitation in the formation. Ferric iron found in a mineralogical form in the formation, however, can be a problem. Also, if particulate metallic iron or iron scale has been carried into the formation, precipitation may occur during treatment. From an operational standpoint, this problem occurs most often in water injection wells.

The precipitation of gelatinous ferric hydroxide can be prevented by adding certain complexing or sequestering agents to the acid. Several organic acids and their derivatives are commonly considered for this application. The most useful organic acids (citric, lactic, acetic, and gluconic acids) and the most effective derivatives [ethylene diamine tetracetic acid (EDTA) and nitrilo triacetic acid (NTA)] were studied in depth by Smith.[25] He found that each material has certain advantages and limitations. Both cost and performance vary widely; performance is affected by both temperature and the presence of other metallic ions. Only citric acid, EDTA, and NTA can hold as much as 3,000 ppm ferric iron in spent acid solutions for more than 4 hours at temperatures above 175°F. Lactic, acetic, and gluconic acids may prevent ferric hydroxide precipitation from spent acid at lower temperatures and iron concentrations or for shorter periods of time. Lactic acid, in particular, is ineffective at bottom-hole temperatures above about 100°F. Smith's studies[25] indicated that several chemicals can be effective complexing agents when used in the correct application. Therefore, there is no one best agent for all applications.

EDTA is one of the most effective agents commonly available because large quantities may be used without precipitation of the calcium salt and it can be used to temperatures of at least 200°F. The major drawback to EDTA is its expense, which often limits its application. Citric acid is one of the least expensive, yet effective, materials. However, when it is used at concentrations greater than 14 lb/1,000 gal, it can precipitate as calcium citrate if less complexable iron is present than that required to react with the quantity of citric acid used. NTA appears to be a useful compromise between EDTA and citric acid because it is more effective than citric acid and less expensive than EDTA. Also, like EDTA, NTA may be used in considerable quantity without precipitation of the calcium salt and is also effective at temperatures up to about 200°F.

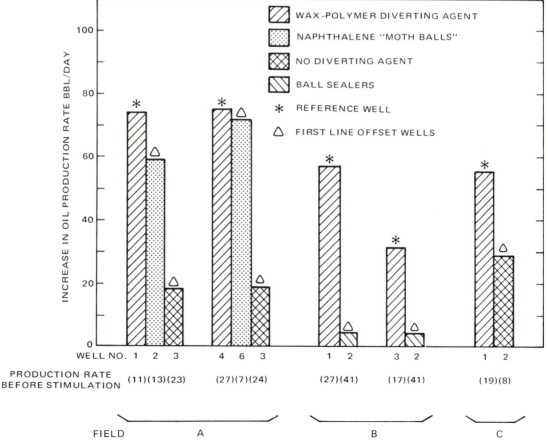

Fig. 11.14—Comparative field results when fracturing with wax-polymer and other diverting agents.[16]

The use of iron-sequestering agents should be considered for water injection wells where corrosion control has been inadequate or where the formation is known to contain ferric iron. The recommended sequestering agent concentration in acid is shown in Table 11.3, adapted from Smith.[25] This tabulation compares agents, showing their advantages and disadvantages and the amount of each required in 1,000 gal of 15-percent hydrochloric acid to sequester 5,000 ppm of ferric iron at 150°F for a minimum of 2 days. If further details are needed about iron-sequestering agents in acidizing treatments, Ref. 25 should be consulted.

11.9 Cleanup Additives

When a problem is anticipated in removing spent acid from the reservoir, the addition of gaseous nitrogen, alcohols, or surfactants to the acid before injection is sometimes considered. These additives are commonly used in low-pressure gas reservoirs.[11,30,31]

Nitrogen is added for two reasons: (1) when the wellbore pressure is reduced, the nitrogen expands and can help to gas lift treating fluid out of well tubulars and (2) if sufficient nitrogen is added, the gas saturation in the reservoir will exceed the saturation at which gas can flow at an earlier time in the cleanup period. Tannich[32] considered these effects and showed how nitrogen addition can greatly speed cleanup in some instances.

To add nitrogen to the treating fluid, liquid nitrogen is brought to the wellsite. The liquid nitrogen is pumped at a controlled rate through a heater, thereby vaporizing the nitrogen. The gaseous nitrogen then is introduced into the acid downstream from the high pressure pumps through an atomizer to finely disperse it in the treating fluid. The pressure required to inject nitrogen into the acid is generated by controlling the rate of heat input.

Alcohols are sometimes added to lower the interfacial tension between the spent acid and formation fluids and to increase the vapor pressure of the spent acid.[33-35] Adding alcohol to acid is most useful in very low-permeability, shaly gas reservoirs, where water is difficult to remove from the rock matrix.

As discussed in Section 11.3, surface active agents also are occasionally added to acid to reduce interfacial tension in an attempt to speed the cleanup process.[29] The use of these additives has diminished in recent years, primarily because absorption of the surfactant or formation solids near the wellbore limits their effectiveness.[5]

References

1. Oakes, B. D.: "Effect of Pressure and Temperature on Hydrochloric Acid Inhibitor Systems," paper presented at the Petroleum Div. of the ACS 137th National Meeting in Cleveland, Ohio, April 5-14, 1960.

2. McDougall, L. A.: "Corrosion Inhibitors for High Temperature Acid Applications," *Materials Protection* (Aug. 1969) **8**, 31-32.

3. Smith, C. F., Dollarhide, F. E., and Byth, N. J.: "Acid Corrosion Inhibitors — Are We Getting What We Need?" *J. Pet. Tech.* (May 1978) 737-747.

4. Woodruff, R. A., Baker, J. R., and Jenkins, R. A., Jr.: "Corrosion Inhibition of Hydrochloric-Hydrofluoric Acid/Mutual Solvent Mixtures at Elevated Temperatures," paper SPE 5645 presented at the SPE-AIME 50th Annual Fall Meeting, Dallas, Sept. 28-Oct. 1, 1975.

5. Hall, B. E.: "The Role of Mutual Solvents in Sandstone Acidizing," *J. Pet. Tech.* (Dec. 1975) 1439-1442.

TABLE 11.3—COMPARISON OF VARIOUS IRON-SEQUESTERING AGENTS

Sequestering Agent	Advantages	Disadvantages	Amount* (lb)
Citric acid	Effective at temperatures of up to 200°F.	Will precipitate as calcium citrate when excess quantities are used.	175
Citric acid-acetic acid mixture	Very effective at lower temperatures.	When indicated amount is used, calcium citrate will precipitate unless at least 2,000 ppm Fe^{+++} is present in spent acid. Efficiency decreases rapidly at temperatures above 150°F.	50 (citric acid) 87 (acetic acid)
Lactic acid	Little chance of calcium lactate precipitation if excessive quantities are used.	Not very effective above 100°F.	190 (at 75°F)
Acetic acid	No problem from possible precipitation as calcium acetate.	Effective only to about 160°F.	435
Gluconic acid	Little chance of calcium gluconate precipitation.	Effective only up to about 150°F. Expensive on a cost-performance basis.	350
Tetrasodium salt of ethylene diamine tetracetic acid (EDTA)	Large quantities may be used without precipitation of calcium salt. Effective at temperatures up to at least 200°F.	More expensive to use than many other agents.	296
Trisodium salt of nitrilo triacetic acid (NTA)	May be used in considerable excess without precipitation as calcium salt. Effective up to at least 200°F.	Less expensive than EDTA, but still more expensive than citric acid.	250

*Amount required in 100 gal of 15-percent HCl acid to sequester 5,000 ppm Fe^{+++} for a minimum 2 days at 150°F.

6. Gidley, J. L.: "Stimulation of Sandstone Formations with the Acid-Mutual Solvent Method," *J. Pet. Tech.* (May 1971) 551-558.

7. Moore, E. W., Crowe, C. W., and Hendrickson, A. R.: "Formation, Effect and Prevention of Asphaltene Sludges During Stimulation Treatments," *J. Pet. Tech.* (Sept. 1965) 1023-1028.

8. Taylor, D. B. and Plummer, R. A.: "Gas Well Stimulation Using Coiled Tubing and Acid with a Mutual Solvent," paper SPE 4115 presented at the SPE-AIME 47th Annual Fall Meeting, San Antonio, Oct. 8-11, 1972.

9. Sutton, G. D. and Lasater, R. M.: "Aspects of Acid Additive Selection in Sandstone Acidizing," paper SPE 4114 presented at the SPE-AIME 47th Annual Fall Meeting, San Antonio, Oct. 8-11, 1972.

10. Dunlap, Peggy M.: "Acidizing Subterranean Formations," U.S. Patent No. 3,254,718 (June 7, 1966).

11. Howard, G. C. and Fast, C. R.: *Hydraulic Fracturing*, Monograph Series, Society of Petroleum Engineers of AIME, Dallas (1970).

12. Harrison, N. W.: "Diverting Agents — History and Application," *J. Pet. Tech.* (May 1972) 593-598.

13. Carpenter, N. F. and Ernst, E. A.: "Acidizing with Swellable Polymers," *J. Pet. Tech.* (Sept. 1962) 1041-1047; *Trans.*, AIME, **225**.

14. Crowe, C. W.: "Evaluation of Oil Soluble Resin Mixtures as Diverting Agents for Matrix Acidizing," paper SPE 3505 presented at the SPE-AIME 46th Annual Fall Meeting, New Orleans, Oct. 3-6, 1971.

15. Nierode, D. E. and Kruk, K. F.: "An Evaluation of Acid Fluid-Loss Additives, Retarded Acids, and Acidized Fracture Conductivity," paper SPE 4549 presented at the SPE-AIME 48th Annual Fall Meeting, Las Vegas, Sept. 30-Oct. 3, 1973.

16. Gallus, J. P. and Pye, D. S.: "Deformable Diverting Agent for Improved Well Stimulation," *J. Pet. Tech.* (April 1969) 497-504; *Trans.*, AIME, **246**.

17. Clark, J. B., Fast, C. R., and Howard, G. C.: "A Multiple-Fracturing Process for Increasing the Productivity of Wells," *Drill. and Prod. Proc.*, API (1952) 104-117.

18. Neill, G. H., Brown, R. W., and Simmons, C. M.: "An Inexpensive Method of Multiple Fracturing," *Drill. and Prod. Proc.*, API (1957) 27.

19. Brown, R. W., Neill, G. H., and Loper, R. G.: "Factors Influencing Optimum Ball Sealer Performance," *J. Pet. Tech.* (April 1963) 450-454.

20. Murphy, W. B. and Juch, A. H.: "Pin-Point Sand-Fracturing — A Method of Simultaneous Injection Into Selected Sands," *J. Pet. Tech.* (Nov. 1960) 21-24.

21. Lagrone, K. W. and Rasmussen, J. W.: "A New Development in Completion Methods — The Limited Entry Technique," *J. Pet. Tech.* (July 1963) 695-702.

22. Williams, B. B., Nieto, G., Graham, H. L., and Leiback, R. E.: "A Staged Fracturing Treatment for Multisand Intervals," *J. Pet. Tech.* (Aug. 1973) 897-904.

23. Gallus, J. P. and Pye, D. S.: "Fluid Diversion to Improve Well Stimulation," paper SPE 3811 presented at the Joint AIME-MMIJ Meeting, Tokyo, May 25-27, 1972.

24. Novotny, E. J.: Private communication, Exxon Production Research Co., Houston.

25. Smith, C. E., Crowe, C. E., and Nocan, T. J. III: "Secondary Deposition of Iron Compounds Following Acidizing Treatments," paper SPE 2358 presented at the SPE-AIME Eastern Regional Meeting, Charleston, W. Va., Nov. 7-8, 1968.

26. Coulter, A. W. and Smithey, C. M.: "Rapid Screening Test Developed To Find Hydrochloric Acid Inhibitors for High Temperature Service," *Materials Protection* (March 1969) **8**, 37.

27. *Corrosion Inhibitors*, Nathan, C. C. (ed.), National Association of Corrosion Engineers, Houston (1973) 156-172.

28. Keeney, B. R., Lasater, R. M. and Knox, J. A.: "New Organic Inhibitor Retards Sulfide Stress Cracking," *Materials Protection* (April 1968) **7**, 23.

29. Grubb, W. E. and Martin, F. G.: "A Guide to Chemical Well Treatments, Part 3: Handy Guide to Treating Chemicals," *Pet. Eng.* (July 1963) 82-86.

30. Neill, G. H., Dobbs, J. B., Pruitt, G. T., and Crawford, H. R.: "Field and Laboratory Results of Carbon Dioxide and Nitrogen in Well Stimulation," *J. Pet. Tech.* (March 1964) 243-248.

31. Foshee, W. C. and Hurst, R. E.: "Improvement of Well Stimulation Fluids by Including a Gas Phase," *J. Pet. Tech.* (July 1965) 768-772.

32. Tannich, J. D.: "Liquid Removal From Hydraulically Fractured Gas Wells," *J. Pet. Tech.* (Nov. 1975) 1309-1317.

33. McLeod, H. O., McGinty, J. E., and Smith, C. F.: "Alcoholic Acid Speeds Clean-up in Sandstones," *Pet. Eng.* (Feb. 1966) 66-70.

34. McLeod, H. O., Smith, C. F., and Ross, W. M.: "Deep Well Stimulation with Alcoholic Acids," paper SPE 1558 presented at the SPE-AIME 41st Annual Fall Meeting, Dallas, Oct. 2-5, 1966.

35. McLeod, H. O. and Coulter, A. W.: "The Use of Alcohol in Gas Well Stimulation," paper SPE 1663 presented at the SPE-AIME Eastern Regional Meeting, Columbus, Ohio (Nov. 1966).

Chapter 12

Acidizing Economics

12.1 Introduction

An evaluation of the economics of a stimulation treatment, once specified, must consider many factors, including: (1) treatment cost, (2) initial increase in production rate, (3) additional reserves that may be produced before the well reaches its economic limit, (4) rate of production decline before and after stimulation, and (5) reservoir and mechanical problems that could cause the treatment to be unsuccessful.

Numerous criteria or ''yardsticks'' can be used to measure the economic worth of a treatment or to compare the relative value of several optional workovers or stimulation treatments. Unfortunately, no single yardstick can adequately satisfy all requirements. Two yardsticks commonly used to analyze stimulation treatments and workovers are the *payout period* and the net *profit-to-investment* ratio. More complicated cash-flow analyses, such as a discounted rate-of-return analysis (DCF), normally are not employed because of the quick payout following a successful stimulation treatment and the very high rate of return that is therefore predicted. The application of DCF methods can be important when expensive acid stimulation treatments are employed, particularly if they are used field wide.

12.2 Cash Flow Analysis

Almost all yardsticks require a knowledge of the net cash flow generated by the stimulation treatment; that is, treatment cost and the net income (after federal income taxes) generated by the treatment. To make this prediction, accurate estimates must be obtained for the following factors: (1) site preparation costs, (2) equipment mobilization and demobilization costs, (3) stimulation treatment cost, (4) current production rate, (5) anticipated production rate after stimulation, (6) production decline rate before and after stimulation, and (7) unit value of incremental production after deduction of operating expenses and federal and state taxes.

To illustrate how these factors are used, consider an acidizing treatment for an oil well completed in a carbonate formation that historically has required repeated treatment to maintain productivity. The well was acidized 2 years ago and averaged 65 BOPD and 10 BWPD the first month after

stimulation. During the past year, production has declined from 35 to 20 BOPD and 8 to 5 BWPD. A recent pressure buildup test indicates the well is damaged and currently is producing at half its undamaged potential.

To establish treatment cost, bids were requested from local service companies. The low service-company charge for equipment mobilization, demobilization, and chemicals totaled $8,500. Additional charges totaling $2,000 are expected to cover pulling and running the rod pump, site preparation, and other miscellaneous items. It is anticipated that this workover will be an expense item. The income-tax rate for the company is 50 percent, so the net treatment cost after federal income taxes will be $5,250.

To estimate income credited to the treatment, an accurate estimate of production rate after stimulation is required. All prior history in the field, or in similar formations, should be used to establish this rate history. *It is seldom realistic to base the analysis only on production rate before and after the treatment.*

The production history for the example well and its expected production after stimulation are plotted in Fig. 12.1. These production curves were predicted assuming (1) the production rate will continue to decline exponentially if the well is not stimulated, (2) the stimulation treatment will successfully remove the effect of near wellbore damage, but will not give any appreciable stimulation other than damage removal, and (3) production after stimulation will decline at the same rate it did following the previous treatment on this well. Note that another stimulation treatment probably will be required in about 18 months, when the production rate reaches 15 BOPD.

12.3 Payout Period

The simplest yardstick for comparing stimulation treatments is the payout period. This is defined as the length of time required for the total cash outlay to be recovered from the increased cash inflow resulting from the treatment. For the example well, the payout period determined from Table 12.1 is 3.25 months if the net after-tax value of the oil is $3/bbl and 1.8 months if the oil value is $5/bbl.

One problem when using the payout period alone as a

TABLE 12.1—SUMMARY OF CASH FLOW FOR EXAMPLE WELL

Time (months)	Item*	Value or Cost (dollars)		Net After Federal Income Tax Cash Flow for the Treatment	
		A	B	A†	B**
0	Stimulation	4,250.	4,250.	−4,250.	−4,250.
0	Other Charges	1,000.	1,000.	−5,250.	−5,250.
1	19 BOPD (570 bbl)	1,710.	2,850.	−3,540.	−2,400.
2	18 BOPD (540 bbl)	1,620.	2,700.	−1,920.	300.
3	17 BOPD (510 bbl)	1,530.	2,550.	−390.	2,850.
4	16 BOPD (480 bbl)	1,440.	2,400.	1,050.	5,250.
5	15 BOPD (450 bbl)	1,350.	2,250.	2,400.	7,500.
6	14 BOPD (420 bbl)	1,260.	2,100.	3,660.	9,600.
7	(400 bbl)	1,200.	2,000.	4,860.	11,600.
8	(380 bbl)	1,140.	1,900.	6,000.	13,500.
9	(360 bbl)	1,080.	1,800.	7,080.	15,300.
10	(340 bbl)	1,020.	1,700.	8,100.	17,000.
11	(320 bbl)	960.	1,600.	9,060.	18,600.
12	10 BOPD (300 bbl)	900.	1,500.	9,960.	20,100.
13	(285 bbl)	855.	1,425.	10,815.	21,525.
14	(270 bbl)	810.	1,350.	11,625.	22,875.
15	(255 bbl)	765.	1,275.	12,390.	24,150.
16	(240 bbl)	720.	1,200.	13,110.	25,350.
17	(225 bbl)	675.	1,125.	13,785.	26,475.
18	7 BOPD (210 bbl)	620.	1,050.	14,415.	27,525.

*Production is the incremental production credited to the stimulation treatment.
**B assumes net after tax value of the incremental oil is $5/bbl.
†A assumes net after tax value of the incremental oil is $3/bbl.

yardstick is that it is not sensitive to the life of the stimulation treatment. For example, if production in the example well was expected to stabilize after 6 months (at 30 BOPD), with the same early time behavior, the payout period would not change. Obviously, the value of the treatment would be increased. Because of this deficiency, the payout period is usually considered along with a profit-to-investment ratio or other parameters.

12.4 Profit-To-Investment Ratio

In its simplest form, the profit-to-investment ratio is defined as

$$F_{pc} = \frac{\text{net profit after federal income tax}}{\text{treatment cost}}.$$

$$\dots\dots\dots\dots\dots\dots\dots\dots\dots\dots (12.1)$$

When money is plentiful and when many opportunities are available, the first year profit-to-investment ratio often is used to develop a priority list for stimulation treatments or workovers. When capital is short, a first year F_{pc} along with an over-all F_{pc}, or a 2-, 3-, or 4-year F_{pc}, may be used to maximize return of capital for future investment. Over-all F_{pc} and first-year F_{pc} ratios for the example treatment are given in Table 12.2.

The risk of failure can be introduced to the payout and F_{pc} calculations to obtain a "risk-weighted" yardstick. Techniques for doing this vary; however, a technique described by Rike* has wide acceptance when evaluating workovers and stimulation treatments. The modified F_{pc} is calculated using Eq. 12.2.

Risk-weighted $F_{pc} =$

$$\frac{\text{cash flow (mechanical risk)(reservoir risk)} - \text{cost}}{\text{cost}}.$$

$$\dots\dots\dots\dots\dots\dots\dots\dots\dots\dots (12.2)$$

*Rike, J. L.: "Workover Economics — Complete but Simple," *J. Pet. Tech.* (Jan. 1972) 67-72.

Rike* proposed we consider two distinct types of risk — a geologic or reservoir risk and a mechanical risk.

● The reservoir risk factor defines the probability that the reserves really are present and can be produced after the treatment.

In the example well, this would be the probability that the well productivity is really one-half the undamaged rate and that the treatment will successfully remove the effects of the damage. Since we were told that a buildup test was conducted and that acid treatments are routinely used successfully in this reservoir, a relatively high probability of success (reservoir risk) of 0.90 will be assumed.

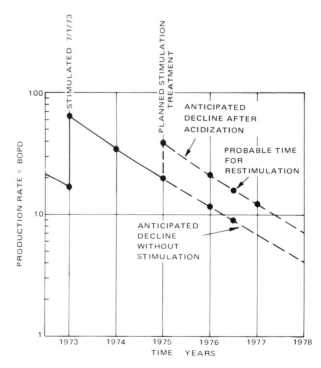

Fig. 12.1—Production history for an example well.

- The mechanical risk defines our uncertainty of being able to mechanically complete the job for the amount of money projected.

A relatively high mechanical risk factor of 0.9 will also be assumed for the example well. Since the service company job costs are fixed, the well tubulars are expected to be in good condition and the well history indicates no unusual mechanical problems.

For the example acid treatment, the risk weighted after 18 months at a profit of $3/bbl therefore would be calculated as follows.

$$\text{Risk-weighted } F_{pc} = \frac{19,665(0.9)(0.9) - 5,250}{5,250},$$

$$\text{Risk-weighted } F_{pc} = 2.03. \quad \dots\dots\dots\dots\dots (12.3)$$

Economic yardsticks for the example treatment for net oil values of $3 and $5/bbl are given in Table 12.2.

The selection of appropriate risk factors is important for economic evaluation. If the factors are too small (pessimistic), the treatment may not be performed; if they are too large (optimistic), uneconomic treatments will be conducted.

There are no easy guidelines for selecting accurate risk factors. These factors depend on the type of treatment, formation characteristics, condition of well tubulars, accuracy of reservoir data, etc. The best procedure for establishing this factor is to rely on statistical data in a given reservoir or reservoir type. When prior data are not available, the choice is of necessity subjective and should reflect the best input from knowledgeable reservoir and production engineers.

12.5 Discounted Cash Flow Techniques

When considering major stimulation projects, it is sometimes necessary to compare alternatives using a yardstick that accurately considers the timing of income and expenditures. A prime example would be the development of a field that is economic only through the application of very large proppant or acid fracturing treatments, different well spacings, etc. To conduct an analysis of this type, it is usually necessary to numerically simulate production from the well before and after stimulation to be sure reservoir effects are accurately considered. Generally, this detail cannot be jus-

TABLE 12.2—ECONOMIC YARDSTICKS FOR THE EXAMPLE WELL

	$3/bbl Net Profit (months)	$5/bbl Net Profit (months)
Payout period	3.25	1.8
Over-all F_{pc}	2.75	5.24
First-year F_{pc}	1.89	3.83
Risk-weighted over-all F_{pc}	2.03	4.06

tified for a single stimulation treatment, unless it is a very high cost treatment.

Cash flow methods give greater weight to income generated in the early years of a project by "discounting" the value of income in later years with a compound interest factor. Probably the most common discounted cash flow yardstick is the discounted cash flow rate, often referred to as a "rate of return." This technique calculates the compound interest rate required to discount all future cash flow so that the sum of the discounted values is equal to the initial cash outlay. An alternative form of this method discounts the cash flow at a standard rate and calculates the excess (or deficiency) of the value of future dollars over the investment.

Because the cash flow techniques are not widely used to evaluate stimulation treatments, this discussion has been prepared to give only an overview. The reader should consult more complete economics texts for details. Also, Rike has documented a short cut technique that can be used in the unlikely event that income is constant following the stimulation treatment.

12.6 Use of Economic Yardsticks

The economic yardsticks just described are best used as comparative tools. Alternative stimulation and workover treatments then can be compared with these yardsticks, and a priority ranking established. The F_{pc} ratio is often used this way. If realistic data have been used to generate the yardsticks, using this priority list to select treatments should maximize return from a stimulation program.

It is beyond the scope of this monograph to discuss absolute criteria for accepting or rejecting a specific stimulation treatment. Each company has established guidelines, based on current economic conditions for its use.

Appendix

Conversion Factors

To convert any physical quantity from one set of units to another, multiply by the appropriate table entry. For example, suppose that p is given as 10 lb-in.$^{-2}$ but is desired in units of poundals-ft^{-2}. From Table A.1, the result is

$$p = 10 \times 4.6330 \times 10^3 \times 4.6330 \times 10^4 \text{ poundals-ft}^{-2}$$

or lbm-ft^{-1}-sec^{-2}.

Besides the extended tables, a few of the commonly used conversion factors are found in the following chart.

To Convert From	To	Multiply by
Pounds	Grams	453.59
Kilograms	Pounds	2.2046
Inches	Centimeters	2.5400
Meters	Inches	39.370
Gallons (U.S.)	Liters	3.7853
Gallons (U.S.)	Cubic inches	231.00
Gallons (U.S.)	Cubic feet	0.13368
Cubic feet	Liters	28.316

TABLE A-1—CONVERSION FACTORS FOR QUANTITIES HAVING DIMENSIONS OF F OR MLt^{-2} (Force)

Given a Quantity in These Units	Multiply by Table Value To Convert to These Units →	G cm sec^{-2} (dynes)	Kg m sec^{-2} (newtons)	Lbm ft sec^{-2} (poundals)	Lbf
G cm sec^{-2}(dynes)		1	10^{-5}	7.2330×10^{-5}	2.2481×10^{-6}
Kg m sec^{-2}(newtons)		10^5	1	7.2330	2.2481×10^{-1}
Lbm ft sec^{-2}(poundals)		1.3826×10^4	1.3826×10^{-1}	1	3.1081×10^{-2}
Lbf		4.4482×10^5	4.4482	32.1740	1

TABLE A-2—CONVERSION FACTORS FOR QUANTITIES HAVING DIMENSIONS OF F/L^2 OR $ML^{-1}t^{-2}$

Given a Quantity in These Units	Multiply by Table Value To Convert to These Units →	G cm^{-1} sec^{-2} (dyne cm^{-2})	Kg m^{-1} sec^{-2} (newtons m^{-2})	Lbm ft^{-1} sec^{-2} (poundals ft^{-2})	Lbf ft^{-2}	Lbf in.$^{-2}$ (psia)*
G cm^{-1} sec^{-2}		1	10^{-1}	6.7197×10^{-2}	2.0886×10^{-3}	1.4504×10^{-5}
Kg m^{-1} sec^{-2}		10	1	6.7197×10^{-1}	2.0886×10^{-2}	1.4504×10^{-4}
Lbm ft^{-1} sec^{-2}		1.4882×10^1	1.4882	1	3.1081×10^{-2}	2.1584×10^{-4}
Lbf ft^{-2}		4.7880×10^2	4.7880×10^1	32.1740	1	6.9444×10^{-3}
Lbf in.$^{-2}$		6.8947×10^4	6.8947×10^3	4.6330×10^3	144	1
Atmospheres		1.0133×10^6	1.0133×10^5	6.8087×10^4	2.1162×10^3	14.696
MM Hg		1.3332×10^3	1.3332×10^2	8.9588×10^1	2.7845	1.9337×10^{-2}
In. Hg		3.3864×10^4	3.3864×10^3	2.2756×10^3	7.0727×10^1	4.9116×10^{-1}

Given a Quantity in These Units	Multiply by Table Value To Convert to These Units →	Atmospheres (atm)	MM Hg	In. Hg
G cm^{-1} sec^{-2}		9.8692×10^{-7}	7.5006×10^{-4}	2.9530×10^{-5}
Kg m^{-1} sec^{-2}		9.8692×10^{-6}	7.5006×10^{-3}	2.9530×10^{-4}
Lbm ft^{-1} sec^{-2}		1.4687×10^{-5}	1.1162×10^{-2}	4.3945×10^{-4}
Lbf ft^{-2}		4.7254×10^{-4}	3.5913×10^{-1}	1.4139×10^{-2}
Lbf in.$^{-2}$		6.8046×10^{-2}	5.1715×10^1	2.0360
Atmospheres		1	760	29.921
MM Hg		1.3158×10^{-3}	1	3.9370×10^{-2}
In. Hg		3.3421×10^{-2}	25.400	1

*This unit is preferably abbreviated psia (pounds per square inch absolute) or psig (pounds per square inch gauge). Gauge pressure is absolute pressure minus the prevailing barometric pressure.

TABLE A-3—CONVERSION FACTORS FOR QUANTITIES HAVING DIMENSIONS OF FL OR ML^3t^2
(Work, Energy, Torque)

Given a Quantity in These Units	Multiply by Table Value To Convert to These Units → G cm² sec⁻² (ergs)	Kg m² sec⁻² (absolute joules)	Lbm ft² sec⁻² (ft-poundals)	Ft lbf	Cal*
G cm² sec⁻²	1	10^{-7}	2.3730×10^{-6}	7.3756×10^{-8}	2.3901×10^{-8}
Kg m² sec⁻²	10^7	1	2.3730×10^1	7.3756×10^{-1}	2.3901×10^{-1}
Lbm ft² sec⁻²	4.2140×10^5	4.2140×10^{-2}	1	3.1081×10^{-2}	1.0072×10^{-2}
Ft lbf	1.3558×10^7	1.3558	32.1740	1	3.2405×10^{-1}
Thermochemical calories*	4.1840×10^7	4.1840	9.9287×10^1	3.0860	1
Btu	1.0550×10^{10}	1.0550×10^3	2.5036×10^4	778.16	2.5216×10^2
Hp-hr	2.6845×10^{13}	2.6845×10^6	6.3705×10^7	1.9800×10^6	6.4162×10^5
Absolute kw-hr	3.6000×10^{13}	3.6000×10^6	8.5429×10^7	2.6552×10^6	8.6042×10^5

Given a Quantity in These Units	Multiply by Table Value To Convert to These Units → Btu	Hp-hr	Kw-hr
G cm² sec⁻²	9.4783×10^{-11}	3.7251×10^{-14}	2.7778×10^{-14}
Kg m² sec⁻²	9.4783×10^{-4}	3.7251×10^{-7}	2.7778×10^{-7}
Lbm ft² sec⁻²	3.9942×10^{-5}	1.5698×10^{-8}	1.1706×10^{-8}
Ft lbf	1.2851×10^{-3}	5.0505×10^{-7}	3.7662×10^{-7}
Thermochemical calories*	3.9657×10^{-3}	1.5586×10^{-6}	1.1622×10^{-6}
Btu	1	3.9301×10^{-4}	2.9307×10^{-4}
Hp-hr	2.5445×10^3	1	7.4570×10^{-1}
Absolute kw-hr	3.4122×10^3	1.3410	1

*This unit, abbreviated cal, is used in chemical thermodynamic tables. To convert quantities expressed in International Steam Table calories (abbreviated I.T. cal) to this unit, multiply by 1.000654.

TABLE A-4—CONVERSION FACTORS FOR QUANTITIES HAVING DIMENSIONS* OF $ML^{-1}t^{-1}$ OR FtL^{-2} OR MOLES $L^{-1}t^{-1}$
(Viscosity, Density times Diffusivity, Concentration times Diffusivity)

Given a Quantity in These Units	Multiply by Table Value To Convert to These Units → G cm⁻¹ sec⁻¹ (poises)	Kg m⁻¹ sec⁻¹	Lbm ft⁻¹sec⁻¹	Lbf sec ft⁻²	Cp	Lbm ft⁻¹ hr⁻¹
G cm⁻¹ sec⁻¹	1	10^{-1}	6.7197×10^{-2}	2.0886×10^{-3}	10^2	2.4191×10^2
Kg m⁻¹ sec⁻¹	10	1	6.7197×10^{-1}	2.0886×10^{-2}	10^3	2.4191×10^3
Lbm ft⁻¹ sec⁻¹	1.4882×10^1	1.4882	1	3.1081×10^{-2}	1.4882×10^3	3,600
Lbf sec ft⁻²	4.7880×10^2	4.7880×10^1	32.1740	1	4.7880×10^4	1.1583×10^5
Cp	10^{-2}	10^{-3}	6.7197×10^{-4}	2.0886×10^{-5}	1	2.4191
Lbm ft⁻¹ hr⁻¹	4.1338×10^{-3}	4.1338×10^{-4}	2.7778×10^{-4}	8.6336×10^{-6}	4.1338×10^{-1}	1

*When moles appear in the given and desired units, the conversion factor is the same as for the corresponding mass units.

TABLE A-5—CONVERSION FACTORS FOR QUANTITIES HAVING DIMENSIONS OF $MLt^{-3}T$ OR $Ft^{-1}T^{-1}$
(Thermal Conductivity)

Given a Quantity in These Units	Multiply by Table Value to Convert to These Units → G cm sec⁻³ °K⁻¹ (ergs sec⁻¹ cm⁻¹ °K⁻¹)	Kg m sec⁻³ °K⁻¹ (watts m⁻¹ °K⁻¹)	Lbm ft sec⁻³ °F⁻¹	Lbf sec⁻¹ °F⁻¹	Cal sec⁻¹ cm⁻¹ °K⁻¹	Btu hr⁻¹ ft⁻¹ °F⁻¹
G cm sec⁻³ °K⁻¹	1	10^{-5}	4.0183×10^{-5}	1.2489×10^{-6}	2.3901×10^{-8}	5.7780×10^{-6}
Kg m sec⁻³ °K⁻¹	10^5	1	4.0183	1.2489×10^{-1}	2.3901×10^{-3}	5.7780×10^{-1}
Lbm ft sec⁻³ °F⁻¹	2.4886×10^4	2.4886×10^{-1}	1	3.1081×10^{-2}	5.9479×10^{-4}	1.4379×10^{-1}
Lbf sec⁻¹ °F⁻¹	8.0068×10^5	8.0068	3.2174×10^1	1	1.9137×10^{-2}	4.6263
Cal sec⁻¹ cm⁻¹ °K⁻¹	4.1840×10^7	4.1840×10^2	1.6813×10^3	5.2256×10^1	1	2.4175×10^2
Btu hr⁻¹ ft⁻¹ °F⁻¹	1.7307×10^5	1.7307	6.9546	2.1616×10^{-1}	4.1365×10^{-3}	1

TABLE A-6—CONVERSION FACTORS FOR QUANTITIES HAVING DIMENSIONS OF L^2t^{-1}
(Momentum Diffusivity, Thermal Diffusivity, Molecular Diffusivity)

Given a Quantity in These Units	Multiply by Table Value To Convert to These Units → Cm² sec⁻¹	M² sec⁻¹	Ft² hr⁻¹	Cs
Cm² sec⁻¹	1	10^{-4}	3.8750	10^2
M² sec⁻¹	10^4	1	3.8750×10^4	10^6
Ft² hr⁻¹	2.5807×10^{-1}	2.5807×10^{-5}	1	2.5807×10^1
Cs	10^{-2}	10^{-6}	3.8750×10^{-2}	1

TABLE A-7—CONVERSION FACTORS FOR QUANTITIES HAVING DIMENSIONS OF $Mt^{-3}T^{-1}$ OR $FL^{-1}t^{-1}T^{-1}$
(Heat-Transfer Coefficients)

Given a Quantity in These Units	Multiply by Table Value To Convert to These Units → $G\ sec^{-3}\,°K^{-1}$	$Kg\ sec^{-3}\,°K^{-1}$ $(W\ m^{-2}\,°K^{-1})$	$Lbm\ sec^{-3}\,°F^{-1}$	$Lbf\ ft^{-1}\ sec^{-1}$ $°F^{-1}$	$Cal\ cm^{-2}sec^{-1}$ $°K^{-1}$
$G\ sec^{-3}\,°K^{-1}$	1	10^{-3}	1.2248×10^{-3}	3.8068×10^{-5}	2.3901×10^{-8}
$Kg\ sec^{-3}\,°K^{-1}$	10^{3}	1	1.2248	3.8068×10^{-2}	2.3901×10^{-5}
$Lbm\ sec^{-3}\,°F^{-1}$	8.1647×10^{2}	8.1647×10^{-1}	1	3.1081×10^{-2}	1.9514×10^{-5}
$Lbf\ ft^{-1}\ sec^{-1}\,°F^{-1}$	2.6269×10^{4}	2.6269×10^{1}	32.1740	1	6.2784×10^{-4}
$Cal\ cm^{-2}\ sec^{-1}\,°K^{-1}$	4.1840×10^{7}	4.1840×10^{4}	5.1245×10^{4}	1.5928×10^{3}	1
$W\ cm^{-3}\,°K^{-1}$	10^{7}	10^{4}	1.2248×10^{4}	3.8068×10^{2}	2.3901×10^{-1}
$Btu\ ft^{-2}\ hr^{-1}\,°F^{-1}$	5.6782×10^{3}	5.6782	6.9546	2.1616×10^{-1}	1.3571×10^{-4}

Given a Quantity in These Units	Multiply by Table Value To Convert to These Units → Watts $cm^{-2}\,°K^{-1}$	$Btu\ ft^{-2}hr^{-1}$ $°F^{-1}$
$G\ sec^{-3}\,°K^{-1}$	10^{-7}	1.7611×10^{-4}
$Kg\ sec^{-3}\,°K^{-1}$	10^{-4}	1.7611×10^{-1}
$Lbm\ sec^{-3}\,°F^{-1}$	8.1647×10^{-5}	1.4379×10^{-1}
$Lbf\ ft^{-1}\ sec^{-1}\,°F^{-1}$	2.6269×10^{-3}	4.6263
$Cal\ cm^{-2}\ sec^{-1}\,°K^{-1}$	4.1840	7.3686×10^{3}
$W\ cm^{-3}\,°K^{-1}$	1	1.7611×10^{3}
$Btu\ ft^{-2}\ hr^{-1}\,°F^{-1}$	5.6782×10^{-4}	1

TABLE A-8—CONVERSION FACTORS FOR QUANTITIES HAVING DIMENSIONS* OF $ML^{-2}t^{-1}$ OR MOLES $L^{-2}t^{-1}$ OR $FL^{-3}t$
(Mass-Transfer Coefficients, k_{xi})

Given a Quantity in These Units	Multiply by Table Value To Convert to These Units → $G\ cm^{-2}\ sec^{-1}$	$Kg\ m^{-2}\ sec^{-1}$	$Lbm\ ft^{-2}\ sec^{-1}$	$Lbf\ ft^{-3}\ sec$	$Lbm\ ft^{-2}\ hr^{-1}$
$G\ cm^{-}\ sec^{-1}$	1	10^{1}	2.0482	6.3659×10^{-2}	7.3734×10^{3}
$Kg\ m^{-2}\ sec^{-1}$	10^{-1}	1	2.0482×10^{-1}	6.3659×10^{-3}	7.3734×10^{2}
$Lbm\ ft^{-2}\ sec^{-1}$	4.8824×10^{-1}	4.8824	1	3.1081×10^{-2}	3,600
$Lbf\ ft^{-3}\ sec$	1.5709×10^{1}	1.5709×10^{2}	32.1740	1	1.1583×10^{5}
$Lbm\ ft^{-2}\ hr^{-1}$	1.3562×10^{-4}	1.3562×10^{-3}	2.7778×10^{-4}	8.6336×10^{-6}	1

*When moles appear in the given and desired units, the conversion factor is the same as for the corresponding mass units.

Nomenclature

Chapter

a_i = activity for molecular species i, moles/volume — 3,4

A = reactive area, area — 4

A = well drainage area, area — 2,7

A = cross-sectional area of a pore, area — 8

c = concentration, moles/volume or mass/volume — 4,8

$\bar{c}$ = average concentration, moles/volume or mass/volume — 4,6,8

c^* = concentration ratio, dimensionless — 4

c_i, c_{io} = concentration of species i and initial concentration of species i, respectively, moles/volume or mass/volume — 3,4,6

c_m = mineral concentration, moles/volume — 8

c_{mi} = concentration of dissoluble minerals not accessible to acid, moles/volume — 8

c_{mo} = initial concentration of dissoluble minerals, moles/volume — 8

c_w = concentration at the wall, moles/volume or mass/volume — 4,6

C = over-all fluid-loss coefficient, length/time$^{1/2}$ — 6,7

C_c = fluid-loss coefficient, compression, length/time$^{1/2}$ — 6

C_v = fluid-loss coefficient, fracture fluid invasion, length/time$^{1/2}$ — 6

C_{vc} = combined fluid-loss coefficient, length/time$^{1/2}$ — 6

C_w = fluid-loss coefficient, filter cake, length/time$^{1/2}$ — 6

D, D_j = diffusion coefficient of component j, length^2/time — 4,6

$D_e, D_e^{(\infty)}$ = effective diffusivity, length^2/time — 4-7

E = Young's modulus, force/area — 5-7

F_{PC} = profit to investment ratio, dimensionless — 12

F_k = permeability ratio, dimensionless — 2

g_f = fracture gradient, force/area/length — 7,9,10

g_{pf} = friction pressure gradient, force/area/length — 7

h = fracture height, length — 5,6

h_g = gross fracture height, length — 2,5-7

h_n = net sand thickness, length — 2,5,7-10

i = injection rate, volume/time — 5-7

i_{max} = maximum injection rate, volume/time — 9,10

J_s = damaged or stimulated formation productivity index, volume/time/length — 2,7

J_o = undamaged formation productivity index, volume/time/length — 2,7

k_{ave} = average permeability, length^2 — 9

Chapter

k = permeability, length^2 — 2,6-10

k_f = fracture permeability, length^2 (see wk_f) — 2,6,7

k_i = initial permeability, length^2 — 8

k_o = original permeability, length^2 — 2

k_s = damaged-zone permeability, length^2 — 2

K = equilibrium constant, various — 3

K_D = dissociation equilibrium constant, various — 3

K_g = mass-transfer coefficient, length/time — 4,6

K_L = dimensionless fracture length, Eq. 6.1, dimensionless — 6

K_{rxn} = equilibrium constant, various — 4

K_s = dimensionless fluid-loss coefficient, Eq. 6.3, dimensionless — 6

K_u = reciprocal dimensionless fracture width, Eq. 6.2, dimensionless — 6

K_{vL} = dimensionless variable, Eq. 6.4, dimensionless — 6

ℓ = length of a pore, length — 4,8

L = fracture or wormhole length, length — 2,4-8

L_{aD} = dimensionless acid penetration distance, $= \dfrac{2xLN_{Re}^*}{\bar{w}N_{Re}}$ dimensionless — 6,7

L_{cD} = dimensionless wormhole length, $= \dfrac{DL}{2\bar{v}_A r_c^2}$, dimensionless — 4,8

L_{fD} = dimensionless fracture length, $= \dfrac{8DL}{3\bar{v}_A \bar{w}^2}$, dimensionless — 4,6

m, m_j = reaction order of component j, dimensionless — 4,6

M_j = moments of the pore-size distribution, area^{j+1} — 8

n = number of pores per unit volume, volume^{-1} — 8

N_{CAP} = acid capacity number $= \dfrac{\phi_o c_o}{\bar{\beta}(1-\phi_o)(c_{mo}-c_{mi})}$, dimensionless — 8

N_{Da} = Damkohler number $= \dfrac{\xi_a(c_{mo}-c_m)L}{u}$, dimensionless — 8

N_{Pe} = Peclet number for flow along fracture $= \dfrac{\bar{w}v_A^{\,o}}{2D_e}$, dimensionless — 6,7

N_{Pe}^* = Peclet number for fluid loss $= \dfrac{\bar{w}v_N}{2D_e}$, dimensionless — 6,7

N_{Re} = Reynolds number for flow down a fracture or a wormhole $= \dfrac{2wv_A^{\,o}\rho}{\mu}$ or $\dfrac{2r_c v_A^{\,o}\rho}{\mu}$, dimensionless — 4,6-8

	Chapter
$N_{Re}*$ = Reynolds number for fluid loss, dimensionless	6-8

$$= \frac{2\overline{w}v_N\rho}{\mu} \text{ (fracture)}$$

$$= \frac{2r_c\overline{v}_N\rho}{\mu} \text{ (wormhole)}$$

	Chapter
N_{Sc} = Schmidt number, $= \frac{\mu}{\rho D_e}$, dimensionless	6,8
N_{Sh} = Sherwood number $= \frac{2\overline{w}\,K_g}{D_e}$, dimensionless	6
p = pressure, force/area	7
Δp = pressure drop, force/area	6,7
$\overline{p}_f$ = average pressure acting to hold fracture open, force/area	6
p_{pc} = pseudo-critical pressure, force/area	7
p_{pr} = psuedo-reduced pressure, dimensionless	7
p_{max} = maximum surface pressure, force/area	9,10
P_c = reaction rate in a wormhole $= \frac{r_c\,\xi'_f\,c_o^{m-1}}{D}$, dimensionless	4,8
P_f = reaction rate in a fracture, $= \frac{\overline{w}\,\xi'_f\,c_o^{m-1}}{2D}$, dimensionless	4,6
q = volumetric flow rate, volume/time	4
r_c = radius of wormhole, length	4,6,8
r_d = drainage radius, length	2,7,9,10
r_i = reaction rate (subscript i dropped when referring to H^+), moles/area/time or mass/area/time	4
r_s = damage-zone radius, length	2
r_w = wellbore radius, length	2,5,7,9,10
$\tilde{r}_{ave}$ = average reaction rate, moles/volume/time	8
$\tilde{rf}(\phi)$ = effective reaction rate, time^{-1}	8
R = gas constant, see Appendix	4
R_c* = extent of reaction in wormhole $= 1-c/c_o$, dimensionless	4,8
R_f* = extent of reaction in fracture $= 1-c/c_o$, dimensionless	4
S_{RE} = rock embedment strength, force/area	6,7
S_g = gas saturation, fraction	7
S_o = oil saturation, fraction	7
S_w = water saturation, fraction	7
t = time, time	4,6-8
t_s = sonic travel time, time/length	7
t_{spt} = spurt-time, Eq. 6.24, time	6
T = temperature, degrees	3-7
T_D = dimensionless temperature, dimensionless	6
T_{inj} = injection temperature, degrees	6
T_{pc} = pseudo critical temperature, degrees	7
T_{pr} = pseudo reduced temperature, dimensionless	7
T_R = reservoir temperature, degrees	6
u = volumetric flux, volume/area-time	4
$u_{A,Y}$ = diffusional flux of component A in Y direction, moles/area/time or mass/area/time	4,7,8
v_A = axial velocity, length/time	4,8
$\overline{v}_A$ = average axial velocity, length/time	4,6
v_N = fluid-loss velocity, length/time	4,6,7
$\overline{v}_N$ = average fluid-loss velocity, length/time	5-7
V = fluid loss to formation, volume/area	6,7

	Chapter
V_f = fracture volume, volume	6,7
V_{inj} = volume injected, volume	7,9
V_R = reactor volume, volume	4
V_{spt} = spurt volume, volume/area	6-9
w = fracture width, length	6
$\overline{w}$ = average fracture width, length	4-7
w_a = dissolved width, Eq. 6.36, length	6,7
w_w = fracture width at the wellbore, length	5,6
wk_f = fracture conductivity, length^3	2,6,7
wk_{fi} = ideal fracture conductivity, length^3	5-7
x = fractional axial position, dimensionless	4-7
x_p = fraction of fracture length that is pressurized, dimensionless	5,6
X = acid dissolving power, volume/volume	3,6,7
X = coordinate position, length	4,8
y = fractional distance perpendicular to center line, dimensionless	4
Y = coordinate position, length	4
α = constant, Eq. 7.2, dimensionless	7-10
β = acid dissolving power, mass/mass	3,8
$\overline{\beta}$ = acid dissolving power (stoichiometric ratio), moles/mole	8
γ = activity coefficient, dimensionless	4,7
Γ = perimeter of a pore, length	8
δ_m = diffusion boundary-layer thickness, length	4
ΔE = activation energy, energy/mole	4
ΔH_{rxn} = heat of reaction, energy/mole	4
η = pore-size distribution, length^{-5}	8
θ = angle related to fractional distance along fracture, dimensionless	6
θ_p = angle related to fracture of fracture length that is pressurized, dimensionless	6
ϕ = angle, Eq. 6.12, radians	6
μ = fluid viscosity, mass/length/time	5-7,9,10
ξ = reaction-rate constant based on chemical activity, various	4,8
ξ_f = forward reaction-rate constant based on chemical activity, various	4,8
ξ_r = reverse reaction-rate constant based on chemical activity, various	4,8
ξ' = reaction-rate constant based on concentration, various	4,6,8
ξ'_a = reaction-rate constant, Eq. 8.20, various	8
ρ = fluid density, mass/volume	3,6-8
ρ_{fl} = formation fluid density, mass/volume	7
ρ_{ma} = formation matrix density, mass/volume	7,8
σ = far field or closure stress, force/area	5-7
τ = reduced time, defined by Eq. 8.14, length	8
ν = Poisson's ratio, dimensionless	6,7
ϕ = porosity, dimensionless	6-8
ϕ_o = original porosity, dimensionless	8
K = isothermal compressibility, (force/area)$^{-1}$	7
K_{fl} = isothermal compressibility of reservoir fluids, (force/area)$^{-1}$	6,7
K_g = isothermal compressibility of gas, (force/area)$^{-1}$	7
K_o = isothermal compressibility of oil, (force/area)$^{-1}$	7
K_{pr} = psuedo reduced compressibility, (force/area)$^{-1}$	7
ψ = pore growth rate, area/time	8

Bibliography

A

Adamson, A. W.: *Physical Chemistry of Surfaces*, Interscience-Wiley, New York (1960).

"A Great Discovery," *Oil City Derrick* (Oct. 10, 1895).

"Acid Treatment Becomes Big Factor in Production," *Oil Weekly* (Oct. 10, 1932) 57.

Applied Engineering Stimulation, BJ-Hughes Inc., Houston.

B

Bancroft, D. H.: "Acid Tests Increase Production of Zwolle Wells," *Oil and Gas J.* (Dec. 22, 1932) 42.

Barenblatt, G. I.: "The Mathematical Theory of Equilibrium Cracks and Brittle Fracture," *Advances in Applied Mechanics*, Academic Press, New York (1962) 7, 55-129.

Baron, G.: "Fracturation hydraulique; bases théoriques, études de laboratoire, essais sur champ," *Proc.*, Seventh World Pet. Cong., Mexico City (1967) 371-393.

Barron, A. N., Hendrickson, A. R., and Wieland, D. R.: "The Effect of Flow on Acid Reactivity in a Carbonate Fracture," *J. Pet. Tech.* (April 1962) 409-415; *Trans.*, AIME, 225.

Beal, Carlton: "The Viscosity of Air, Water, Natural Gas, Crude Oil and its Associated Gases at Oil Field Temperatures and Pressures," *Trans.*, AIME (1946) 165, 103-115.

Bird, R. B., Stewart, W. E., and Lightfoot, E. N.: *Transport Phenomena*, John Wiley & Sons, Inc., New York (1960).

Blumberg, A. A.: "Differential Thermal Analysis and Heterogeneous Kinetics: The Reaction of Vitreous Silica with Hydrofluoric Acid," *J. Phys. Chem.* (1959) 63, 1129.

Blumberg, A. A. and Stavrinou, S. C.: "Tabulated Functions for Heterogeneous Reaction Rates: The Attack of Vitreous Silica by Hydrofluoric Acid," *J. Phys. Chem.* (1960) 64, 1438.

Bombardieri, C. C. and Martin, T. H.: "Acid Treating Process," U.S. Patent No. 3,251,415 (May 17, 1966).

Boomer, D. R., McCune, C. C., and Fogler, H. S.: "Rotating Disk Apparatus for Studies in Corrosive Liquid Environments," *Review Scientific Instruments* (1972) 43, 225.

Broaddus, G. C. and Knox, J. A.: "Influence of Acid Type and Quantity in Limestone Etching," paper API 851-39-I presented at the API Mid-Continent Meeting, Wichita, Kan., March 31-April 2, 1965.

Brown, R. W., Neill, G. H., and Loper, R. G.: "Factors Influencing Optimum Ball Sealer Performance," *J. Pet. Tech.* (April 1963) 450-454.

Burdine, N. T., Gournay, L. S., and Reichertz, P. P.: "Pore Size Distribution of Petroleum Reservoir Rocks," *Trans.*, AIME (1950) 189, 195-204.

C

Cannon, G. E.: "Mud Acid and Formation Washing Agents," Topic No. 31-B, Standard Oil (New Jersey) special reports (1942).

Carpenter, N. F. and Ernst, E. A.: "Acidizing with Swellable Polymers," *J. Pet. Tech.* (Sept. 1962) 1041-1047; *Trans.*, AIME, 225.

Chamberlain, L. C. and Boyer, R. F.: "Acid Solvents for Oil Wells," *Ind. and Eng. Chem.* (1939) 31, 400-406.

Chang, C. Y., Guin, J. A., and Roberts, L. D.: "Surface Reaction with Combined Forced and Free Convection," *AIChE Jour.* (1976) 22, 252.

Chapman, M. E.: "Some of the Theoretical and Practical Aspects of the Acid Treatment of Limestone Wells," *Oil and Gas J.* (Oct. 12, 1933) 10.

Chatelain, J. C., Silberberg, I. H., and Schechter, R. S.: "Thermodynamic Limitations in Organic Acid-Limestone Systems," *Soc. Pet. Eng. J.* (Aug. 1970) 189-195.

"Chemical Company Forms Company to Treat Wells," *Oil Weekly* (Nov. 28, 1932) 59.

"Chemical Treatment Halts Junking Breckenridge Wells," *Oil Weekly* (Feb. 13, 1932) 40.

Chew, Ju-Nan and Connally, C. A., Jr.: "A Viscosity Correlation for Gas Saturated Crude Oils," *Trans.*, AIME (1959) 216, 23-25.

Clark, J. B., Fast, C. R., and Howard, G. C.: "A Multiple-Fracturing Process for Increasing the Productivity of Wells," *Drill. and Prod. Prac.*, API (1952) 104-117.

Clason, C. E. and Staudt, J. G.: "Limestone Reservoir Rocks of Kansas React Favorably to Acid Treatment," *Oil and Gas J.* (April 25, 1935) 53.

Claycomb, J. R. and Schweppe, J. L.: "A Numerical Method for Determining Circulation and Injection Temperatures," paper 67-PET-3 presented at the ASME Petroleum Mechanical Engineering Conference, Philadelphia, Sept. 17-20, 1967.

Cooke, C. E., Jr.: "Conductivity of Fracture Proppants in Multiple Layers," *J. Pet. Tech.* (Sept. 1973) 1101-1197; *Trans.*, AIME, 255.

Corrosion Inhibiters, Nathan, C. C.(ed.) National Assn. Corrosion Engineers, Houston (1973) 156-172.

Coulter, A. W. and Smithey, C. M.: "Rapid Screening Test Developed To Find Hydrochloric Acid Inhibitors for High Temperature Service," *Materials Protection* (March 1969) **8,** 37.

Crowe, C. W.: "Evaluation of Oil Soluble Resin Mixtures as Diverting Agents for Matrix Acidizing," paper SPE 3505 presented at the SPE-AIME 46th Annual Fall Meeting, New Orleans, Oct. 3-6, 1971.

D

Daneshy, A. A.: "On the Design of Vertical Fractures," *J. Pet. Tech.* (Jan. 1973) 83-97; *Trans.,* AIME, **255.**

Day, F. H.: *The Chemical Elements in Nature,* Reinhold Publishing Co., New York (1963) 61.

deGroot, S. R. and Mazur, P.: *Non-Equilibrium Thermodynamics,* North-Holland Publishing Co., Amsterdam (1962) 367-375.

Denbigh, K. G.: *The Principles of Chemical Equilibrium,* Cambridge University Press (1961) 268-276.

Dilgren, R. E.: "Acidizing Oil Formations," U.S. Patent No. 3,215,199 (Nov. 2, 1965).

Dilgren, R. E.: "Acidizing Oil Formations," U.S. Patent No. 3,297,090 (Jan. 10, 1967).

Dilgren, R. E. and Neuman, F. M.: "Acidizing Oil Formations," U.S. Patent No. 3,307,630 (March 7, 1967).

Dill, W. R.: "Reaction Times of Hydrochloric-Acetic Acid Solution on Limestone," paper presented at the ACS 16th Southwest Regional Meeting, Oklahoma City, Dec. 1-3, 1960.

Dorsey, N. E.: "Properties of Ordinary Water Substances in all 'Its' Phases," *Monograph Series No. 81,* ACS (1940).

"Dow Chemical Company v. Halliburton Oil Well Cementing Company," Opinion of Circuit Court of Appeals, Sixth Circuit, *U.S. Patent Quarterly,* 90.

Dullien, F. A. L.: "Pore Structure Analysis" paper presented at the ACS State-of-the-Art Summer Symposium, Div. Ind. Eng. Chem., Washington, D.C. (June 1969).

Dunlap, Peggy M.: "Acidizing Subterranean Formations," U.S. Patent No. 3,254,718 (June 7, 1966).

Dunlop, P.: "Mechanisms of the HCl-CaCO₃ Reaction," paper presented at the ACS 17th Southwest Meeting, Dec. 1961.

Dunn, L. A. and Stokes, R. H.: "Diffusion Coefficients of Monocarboxylic Acids at 25°C," *Aust. J. Chem.* (1965) **18,** 285.

Dysart, G. R. and Whitsitt, N. F.: "Fluid Temperatures in Fractures," paper SPE 1902 presented at the SPE-AIME 42nd Annual Fall Meeting, Houston, Oct. 1-4, 1967.

E

Eickmeier, J. R. and Ramey, H. J., Jr.: "Wellbore Temperature and Heat Losses During Production or Injection Operations," paper 7016 presented at the CIM 21st Annual Technical Meeting, Calgary, May 6-8, 1970.

Engineering Data Book, Natural Gas Processors Suppliers Assn., Tulsa (1977).

F

Farley, J. T., Miller, B. M., and Schoettle, W.: "Design Criteria for Matrix Stimulation for Hydrochloric-Hydrofluoric Acid," *J. Pet. Tech.* (April 1970) 433-440.

Fatt, I. and Chittum, J. F.: "Method of Treating Wells with Acid," U.S. Patent No. 2,910,436 (Oct. 27, 1959).

Finkle, P., Draper, H. D., and Hildebrand, J. H.: "The Theory of Emulsification," *Jour.,* ACS (1923) **45,** 2780.

Fitzgerald, P. E.: "A Review of the Chemical Treatment of Wells," *J. Pet. Tech.,* (Sept. 1953) 11-13.

Fitzgerald, P. E.: "For Lack of a Whale," (unpublished notes on a history of Dowell) Dowell, Houston.

Flood, H. L.: "Current Developments in the Use of Acids and Other Chemicals in Oil Production Problems," *Pet. Eng.* (Oct. 1940) 46.

Fogler, H. S., Lund, K., McCune, C. C., and Ault, J. W.: "Dissolution of Selected Minerals in Mud Acid," paper 52c presented at the AIChE 74th National Meeting, New Orleans, March 11-15, 1973.

Fogler, H. S., Lund, K., McCune, C. C., and Ault, J. W.: "Kinetic Rate Expressions for Reactions of Selected Minerals with HCl and HF Mixtures," paper SPE 4348 presented at the SPE-AIME Oilfield Chemistry Symposium, Denver, May 24-25, 1973.

Foshee, W. C., and Hurst, R. E.: "Improvement of Well Stimulation Fluids by Including a Gas Phase," *J. Pet. Tech.* (July 1965) 768-772.

Frac Guide Data Book, Dowell, Tulsa.

The Fracbook Design/Data Manual, Halliburton Services, Ltd., Houston.

Frasch, H.: "Increasing the Flow of Oil Wells," U.S. Patent No. 556, 669 (March 17, 1896).

Fredrickson, S. E. and Broaddus, G. C.: "Selective Placement of Fluids in a Fracture by Controlling Density and Viscosity," *J. Pet. Tech.* (May 1976) 597-602.

G

Gallus, J. P. and Pye, D. S.: "Deformable Diverting Agent for Improved Well Stimulation," *J. Pet. Tech.* (April 1969) 497-504; *Trans.,* AIME, **246.**

Gallus, J. P. and Pye, D. S.: "Fluid Diversion To Improve Well Stimulation," paper SPE 3811 presented at the Joint AIME-MMIJ Meeting, Tokyo, May 25-27, 1972.

Garland, C. W., Tong, S., and Stockmayer, W. H.: "Diffusion of Chloroacetic Acid in Water," *J. Phys. Chem.* (1965) **69,** 2469.

Gatewood, J. R., Hall, B. E., Roberts, L. D., and Lasater, R. M.: "Predicting Results of Sandstone Acidizing," *J. Pet. Tech.* (June 1970) 693-700.

Geertsma, J. and de Klerk, F.: "A Rapid Method of Predicting Width and Extent of Hydraulically Induced Fractures," *J. Pet. Tech.* (Dec. 1969) 1571-1581; *Trans.,* AIME, **246.**

Gidley, J. L.: "Method for Acid Treating Carbonate Formations," U.S. Patent No. 3,441,085 (April 29, 1969).

Gidley, J. L.: "Stimulation of Sandstone Formations with the Acid-Mutual Solvent Method," *J. Pet. Tech.* (May 1971) 551-558.

Gidley, J. L. and Hanson, H. R.: "Central-Terminal Upset From Well Treatment Is Prevented," *Oil and Gas J.* (Feb. 11, 1974) 53-55.

Gidley, J. L., Ryan, T. P., and Mayfield, T. H.: "A Field Study of Sandstone Acidizing," paper SPE 5693 presented at the SPE-AIME Second Formation Damage Symposium, Houston, Jan. 29-30, 1976.

Gidley, J. L. and Tomer, F. S.: "Acid Fracturing Process," U.S. Patent No. 3,451,818 (July 1, 1969).

Gill, W. N.: "Convective Diffusion in Laminar and Turbulent Reverse Osmosis Systems," *Surface and Colloid Science,*

114

John Wiley & Sons, Inc., New York (1971) **4**, 263.

Gilliland, E. R., Baddour, R. F., and Goldstein, D. J.: "Counter Diffusion of Ions in Water," *Cdn. J. Chem. Eng.* (1957) **37**, 10.

Glover, M. C. and Guin, J. A.: "Permeability Changes in a Dissolving Porous Medium," *AIChE Jour.* (Nov. 1973) **19**, 1190-1195.

Goldsmith, J. R.: *Researches in Geochemistry,* John Wiley & Sons, Inc., New York (1959) 337.

Graham, J. W., Kerver, J. K., and Morgan, F. A.: "Method of Acidizing and Introducing a Corrosion Inhibition Into a Hydrocarbon Producing Formation," U.S. Patent No. 3,167,123 (Jan. 26, 1965).

Gregory, D. P. and Riddiford, A. C.: "Transport to the Surface of a Rotating Disc," *J. Chem. Soc.* (1956) 3756.

Grubb, W. E. and Martin, F. G.: "A Guide to Chemical Well Treatments, Part 3: Handy Guide to Treating Chemicals," *Pet. Eng.* (July 1963) 82-86.

Guin, J. A.: "Chemically Induced Changes in Porous Media," PhD dissertation, U. of Texas (Jan. 1970); also Report No. UT 69-2, Texas Petroleum Research Committee, Austin (1969).

Guin, J. A. and Schechter, R. S.: "Matrix Acidization with Highly Reactive Acids," *Soc. Pet. Eng. J.* (Sept. 1971) 390-398; *Trans.,* AIME, **251**.

Guin, J. A., Silberberg, I. H., and Schechter, R. S.: "Chemically Induced Changes in Porous Media," *Ind. Eng. Chem. Fund.* (1971) **10**, 50-54.

H

Hall, B. E.: "The Role of Mutual Solvents in Sandstone Acidizing," *J. Pet. Tech.* (Dec. 1975) 1439-1442.

Ham, William E.: *Classification of Carbonate Rocks,* Memoir 1, AAPG, Tulsa (1962).

Handbook of Chemistry and Physics, 48th ed., The Chemical Rubber Co., Cleveland, Ohio (1967).

Handbook of Tables for Applied Engineering Science, Chemical Rubber Co., Cleveland, Ohio (1970).

Harned, H. S. and Owens, B.: *Physical Chemistry of Electrolytic Solutions,* 3rd ed., Reinhold Publishing Co., New York (1958).

Harris, F. N.: "Applications of Acetic Acid to Well Completion, Stimulation, and Reconditioning," *J. Pet. Tech.* (July 1961) 637-639.

Harrison, N. W.: "A Study of Stimulation Results of Ellenberger Gas Wells in the Delaware-Val Verde Basins," *J. Pet. Tech.* (Aug. 1967) 1017-1021.

Harrison, N. W.: "Diverting Agents — History and Application," *J. Pet. Tech.* (May 1972) 593-598.

Hathorn, D. H.: personal communication, Halliburton Services, Duncan, Okla. (June 7, 1971).

Hendrickson, A. R., Rosene, R. B., and Wieland, D. R.: "Acid Reaction Parameters and Reservoir Characteristics Used in the Design of Acid Treatments," paper presented at the ACS 137th National Meeting, Cleveland, Ohio, April 5-14, 1960.

Herrington, C. G.: "Recent Developments in the Chemical Treatment of Wells," Topic No. 31-C, Standard Oil (New Jersey) special reports (1942).

Hidy, R. W. and Hidy, M. E.: "Pioneering in Big Business — History of Standard Oil Company (New Jersey) 1882-1911," Harper & Brothers, New York (1955) 156.

Hill, W. L. and Wahl, H. A.: "The Effect of Formation Temperature on Hydraulic Fracture Design," paper 851-42-1 presented at the API Spring Meeting Mid-Continent Div., Amarillo, Tex., April 1968.

Howard G. C. and Fast, C. R.: *Hydraulic Fracturing,* Monograph Series, Society of Petroleum Engineers of AIME, Dallas (1970).

Howard, G. C. and Fast, C. R.: "Optimum Fluid Characteristics for Fracture Extension," *Drill. and Prod. Prac.,* API (1957) 261-270.

I

International Critical Tables, McGraw-Hill Book Co., Inc., New York (1926).

International Critical Tables, McGraw-Hill Book Co., Inc., New York (1933) 233.

J

Jazraui, Waleed: private communication, Imperial Oil Ltd., Calgary.

Joyner, H. D. and Lovingfoss, W. J.: "Use of a Computer Model in Matching History and Predicting Performance of Low Permeability Gas Wells," *J. Pet. Tech.* (Dec. 1971) 1415-1420.

K

Keeney, B. R., Lasater, R. M., and Knox, J. A.: "New Organic Inhibitor Retards Sulfide Stress Cracking, *Materials Protection* (April 1968) **7**, 23.

Keller, H. H., Couch, E. J., and Berry, P. M.: "Temperature Distribution in Circulating Mud Columns," *Soc. Pet. Eng. J.* (Feb. 1973) 23-30.

Kiel, O. M.: "A New Hydraulic Fracturing Process," *J. Pet. Tech.* (Jan. 1970) 89-96.

Kiel, O. M. and Weaver, R. H.: "Emulsion Fracturing System," *Oil and Gas J.* (Feb. 1972) **70**, 72-73.

Klock, G. O.: "Pore Size Distributions as Measured by the Mercury Intrusion Method and Their Use in Predicting Permeability," PhD thesis, Oregon State U., Corvallis (1968).

Knox, T. A., Pollock, R. W., and Beecroft, W. H.: "The Chemical Retardation of Acid and How It Can Be Utilized," *J. Cdn. Pet. Tech.* (Jan.-March 1965) 5-12.

Khristianovich, S. A. and Zheltov, J. P.: "Formation of Vertical Fractures by Means of a Highly Viscous Liquid," *Proc.,* Fourth World Pet. Cong., Rome (1955) 579-586.

Kucera, C. H.: "New Oil Gelling Systems Prevent Damage in Water Sensitive Sands," paper SPE 3503 presented at the SPE-AIME 46th Annual Fall Meeting, New Orleans, Oct. 3-6, 1971.

L

Labrid, J. C.: "Stimulation chemique — etude théorique et expérimentale des équilibres chimiques décrivant l'attaque fluorhydrique d'un grès argileux," *Revue* (Oct. 1971) L'Institut Francais du Petrole, **26**, 855.

Labrid, J. C.: "Thermodynamics and Kinetics Aspects of Argillaceous Sandstone Acidizing," *Soc. Pet. Eng. J.* (April 1975) 117-128.

Lagrone, K. W. and Rasmussen, J. W.: "A New Development in Completion Methods — The Limited Entry Technique," *J. Pet. Tech.* (July 1963) 695-702.

Laidler, K. J.: *Chemical Kinetics,* 2nd. ed., McGraw-Hill Book Co., Inc., New York (1965) 286-296.

Lenhard, P. J.: "Mud Acid — Its Theory and Application to Oil and Gas Wells," *Pet. Eng.*, Annual Issue (1943) 82-89.

Leutwyler, K.: "Casing Temperature Studies in Steam Injection Wells," paper SPE 1264 presented at the SPE-AIME Annual Fall Meeting, Denver, Oct. 3-6, 1965.

Levich, V. G.: *Physiochemical Hydrodynamics*, Prentice-Hall, Inc., Englewood Cliffs, N.J. (1962) 60-75.

Lewis, G. N. and Randall, M. (revised by Pitzer, K. S. and Brewer, L.): *Thermodynamics*, McGraw-Hill Book Co., Inc., New York (1961) Chap. 15.

Lund, K.: "On the Acidization of Sandstone," PhD thesis, U. of Michigan, Ann Arbor (1974).

Lund, K., Fogler, H. S., and McCune, C. C.: "Acidization I: The Dissolution of Dolomite in Hydrochloric Acid," *Chem. Eng. Sci.* (1973) **28**, 681-700.

Lund, K., Fogler, H. S., McCune, C. C., and Ault, J. W.: "Acidization II: The Dissolution of Calcite in Hydrochloric Acid," *Chem. Eng. Sci.* (1975) **30**, 825-835.

Lund, K., Fogler, H. S., and McCune, C. C.: "On Predicting the Flow and Reaction of HCl/HF Acid Mixtures in Porous Sandstone Cores," *Soc. Pet. Eng. J.* (Oct. 1976) 248-260; *Trans.*, AIME, **261**.

Lybarger, J. H., Schenerman, R. F., and Karnes, G. T.: "Buffer-Regulated Treating Fluid Positioning Process," U.S. Patent No. 3,868,996 (1975).

M

McCann, B. E.: "Chemistry Dons a Hard Hat . . . The Saga of Acidizing Service," *Drilling* (July 1968) **29**, 35.

McCune, C. C., Fogler, J. S., Lund, K., Cunningham, J. R., and Ault, J. W.: "A Model for the Physical and Chemical Changes in Sandstone During Acidizing," paper SPE 5157 presented at the SPE-AIME Annual Fall Technical Conference and Exhibition, Dallas, Sept. 28-Oct. 1, 1975.

McCune, C. C., Fogler, J. S., Lund, K., Cunningham, J. R., and Ault, J. W.: "A New Model of the Physical and Chemical Changes in Sandstone During Acidizing," *Soc. Pet. Eng. J.* (Oct. 1975) 361-370.

McDougall, L. A.: "Corrosion Inhibitors for High Temperature Acid Applications," *Materials Protection* (Aug. 1969) **8**, 31-32.

McGuire, W. J. and Sikora, V. J.: "The Effect of Vertical Fractures on Well Productivity," *Trans.*, AIME (1960) **219**, 401-403.

McLeod, H. O. and Coulter, A. W.: "The Use of Alcohol in Gas Well Stimulation," paper SPE 1663 presented at the SPE-AIME Eastern Regional Meeting, Columbus, Ohio, Nov. 11-12, 1966.

McLeod, H. O., McGinty, J. E., and Smith, C. F.: "Alcoholic Acid Speeds Clean-Up in Sandstones," *Pet. Eng.* (Feb. 1966) 66-70.

McLeod, H. O., Smith, C. F., and Ross, W. M.: "Deep Well Stimulation with Alcoholic Acids," paper SPE 1558 presented at the SPE-AIME 41st Annual Fall Meeting, Dallas, Oct. 2-5, 1966.

Miller, D. G.: "Application of Irreversible Thermodynamics to Electrolyte Solutions I: Determination of Ionic Transport Coefficients ℓ_{ij} for Isothermal Vector Transport Processes in Binary Electrolyte Systems," *J. Phys. Chem.* (1966) **70**, 2639.

Miller, D. G.: "Application of Irreversible Thermodynamics to Electrolyte Solutions II: Ionic Coefficients ℓ_{ij} for Isothermal Vector Transport Processes in Ternary Systems," *J. Phys. Chem.* (1967) **71**, 66.

Moore, E. W., Crowe, C. W., and Hendrickson, A. R.: "Formation, Effect and Prevention of Asphaltene Sludges During Stimulation Treatments," *J. Pet. Tech.* (Sept. 1965) 1023-1028.

Moore, W. W.: "Acid Treatment Proved Beneficial to North Louisiana Gas Wells," *Oil Weekly* (Oct. 29, 1934) 31.

Morian, S. C.: "Removal of Drilling Mud from Formation by Use of Acid," *Pet. Eng.* (May 1940) 117.

Morse, R. A. and Von Gonten, W. D.: "Productivity of Vertically Fractured Wells Prior to Stabilized Flow," *J. Pet. Tech.* (July 1973) 807-811.

Moss, J. T. and White, P. D.: "How to Calculate Temperature Profiles in Water Injection Wells," *Oil and Gas J.* (March 9, 1959) 174-178.

Mowrey, S. L.: "The Theory of Matrix Acidization and the Kinetics of Quartz-Hydrogen Fluoride Acid Reactions," MS thesis, U. of Texas, Austin (1974). Also available as Report No. UT 73-4, Texas Petroleum Research Committee, Austin (1974).

"Mud Acid," *The Acidizer*, Dowell Inc. (June 1940) **4**, No. 3.

Murphy, W. B. and Juch, A. A.: "Pin-Point Sand-Fracturing — A Method of Simultaneous Injection Into Selected Sands," *J. Pet. Tech.* (Nov. 1960) 21-24.

Muskat, M.: *Physical Principles of Oil Production*, McGraw-Hill Book Co., Inc., New York (1949) 242.

N

Neill, G. H., Brown, R. W., and Simmons, C. M.: "An Inexpensive Method of Multiple Fracturing," *Drill. and Prod. Prac.*, API (1957) 27.

Neill, G. H., Dobbs, J. B., Pruitt, G. T., and Crawford, H. R.: "Field and Laboratory Results of Carbon Dioxide and Nitrogen in Well Stimulation," *J. Pet. Tech.* (March 1964) 243-248.

Newcombe, R. B.: "Acid Treatment for Increasing Oil Production," *Oil Weekly* (May 29, 1933) 19.

Nierode, D. E.: private communication, Exxon Production Research Co., Houston.

Nierode, D. E., Bombardieri, C. C., and Williams, B. B.: "Prediction of Stimulation from Acid Fracturing Treatments," *J. Cdn. Pet. Tech.* (Oct.-Dec. 1972) 31-41.

Nierode, D. E. and Kruk, K. F.: "An Evaluation of Acid Fluid-Loss Additives, Retarded Acids, and Acidized Fracture Conductivity," paper SPE 4549 presented at the SPE-AIME 48th Annual Fall Meeting, Las Vegas, Sept. 30-Oct. 3, 1973.

Nierode, D. E. and Williams, B. B.: "Characteristics of Acid Reactions in Limestone Formations," *Soc. Pet. Eng. J.* (Dec. 1971) 406-418; *Trans.*, AIME, **251**.

Nierode, D. E., Williams, B. B., and Bombardieri, C. C.: "Prediction of Stimulation from Acid Fracturing Treatments," *J. Cdn. Pet. Tech.* (Oct.-Dec. 1972) 31-41.

Nordgren, R. P.: "Propagation of a Vertical Hydraulic Fracture," *Soc. Pet. Eng. J.* (Aug. 1972) **12**, 306-314.

"North Louisiana Operators Pleased with Acid Results," *Oil Weekly* (Dec. 19, 1932) 70.

Notter, R. H. and Sleicher, C. A.: "The Eddy Diffusivity in the Turbulent Boundary Layer Near a Wall," *Chem. Eng. Sci.* (1972) 2073-2093.

Novotny, E. J.: "Prediction of Stimulation from Acid Fracturing Treatments Using Finite Fracture Conductivity," paper SPE 6123 presented at the SPE-AIME 50th Annual Fall Technical Conference and Exhibition, New Orleans, Oct. 3-6, 1976.

Novotny, E. J.: private communication, Exxon Production Research Co., Houston.

O

Oakes, B. D.: "Effect of Pressure and Temperature on Hydrochloric Acid Inhibitor Systems," paper presented at the Petroleum Div., ACS 137th National Meeting, Cleveland, Ohio, April 5-14, 1960.

P

Perkins, T. K. and Kern, L. R.: "Width of Hydraulic Fracture," *J. Pet. Tech.* (Sept. 1961) 937-949.

Perkins, T. K. and Krech, W. W.: "The Energy Balance Concept of Hydraulic Fracturing," *Soc. Pet. Eng. J.* (March 1968) 1-12.

Perrine, R. L.: "Method of Acidizing Petroliferous Formations," U.S. Patent No. 2,863,832 (Dec. 9, 1958).

Perry, J. H.: *Chemical Engineers Handbook,* 3rd ed., McGraw-Hill Book Co., Inc., New York (1950).

Pickering, S. U.: "Emulsions," *J. Chem. Soc.* (1907) 2001.

Pitzer, P. W. and West, C. K.: "Acid Treatment of Lime Wells Explained and Methods Described," *Oil and Gas J.* (Nov. 22, 1934) 38.

Prats, M.: "Effect of Vertical Fractures on Reservoir Behavior — Incompressible Fluid Case," *Soc. Pet. Eng. J.* (June 1961) 105-118; *Trans.,* AIME, **222.**

Prats, M., Hazebroek, P., and Strickler, W. R.: "Effect of Vertical Fractures on Reservoir Behavior — Compressible Fluid Case," *Soc. of Pet. Eng. J.* (June 1962) 87-94; *Trans.,* AIME, **225.**

Product Technical Guide, The Western Co., Fort Worth.

Purcell, W. R.: "Capillary Pressures — Their Measurement Using Mercury and the Calculation of Permeability Thereof," *Trans.,* AIME (1949) **186,** 39-48.

Putnam, S.: "The Dowell Process to Increase Oil Production," *Ind. Eng. Chem.* (Feb. 20, 1933) 51.

Putnam, S. W.: "Development of Acid Treatment of Oil Wells Involves Careful Study of Problems of Each," *Oil and Gas J.* (Feb. 23, 1933) 8.

Putnam, S. W. and Fry, W. A.: "Chemically Controlled Acidization of Oil Wells," *Ind. Eng. Chem.* (1934) **26,** 921.

R

Ramey, H. J., Jr.: "Wellbore Heat Transmission," *J. Pet. Tech.* (April 1962) 427-435.

Raymond, L. R.: "Temperature Distribution in a Circulating Drilling Fluid," *J. Pet. Tech.* (March 1969) 333-341; *Trans.,* AIME, **246.**

Raymond, L. R. and Binder, G. G.: "Productivity of Wells in Vertically Fractured, Damaged Formations," *J. Pet. Tech.* (Jan. 1967) 120-130; *Trans.,* AIME, **240.**

Resibois, P. M. V.: *Electrolyte Theory,* Harper and Row, New York (1968) Chap. 3.

Rike, J. L.: "Workover Economics — Complete but Simple," *J. Pet. Tech.* (Jan. 1972) 67-72.

Roberts, H.: "Creative Chemistry — A History of Halliburton Laboratories 1930-1958," Halliburton Oil Well Cementing Co., Duncan, Okla. (Jan. 7, 1959).

Roberts, L. D. and Guin, J. A.: "A New Method for Predicting Acid Penetration Distance," *Soc. Pet. Eng. J.* (Aug. 1975) 277-286.

Roberts, L. D. and Guin, J. A.: "The Effect of Surface Kinetics in Fracture Acidizing," *Soc. Pet. Eng. J.* (Aug. 1974) 385-395; *Trans.,* AIME, **257.**

Robinson, R. A. and Stokes, R. H.: *Electrolyte Solutions,* Butterworths, London (1965) 517.

Rose, W. D. and Bruce, W. A.: "Evaluation of Capillary Character in Petroleum Reservoir Rock," *Trans.,* AIME (1949) **186,** 127-142.

Rowan, G.: "Theory of Acid Treatments of Limestone Formations," *I. Inst. Petrol.* (1957) **45,** No. 431.

Russell, D. G. and Truitt, N. E.: "Transient Pressure Behavior in Vertically Fractured Reservoirs," *J. Pet. Tech.* (Oct. 1964) 1159-1170; *Trans.,* AIME, **231.**

S

Sarmiento, R.: "Geological Factors Influencing Porosity Estimates from Velocity Logs," *Bull.,* AAPG (May 1961) **45,** No. 5, 633-644.

Schechter, R. S. and Gidley, J. L.: "The Change in Pore Size Distribution From Surface Reactions in Porous Media," *AIChE Jour.* (May 1969) **15,** 339-350.

Scheidegger, A. E.: *The Physics of Flow Through Porous Media,* The Macmillan Co., New York (1960).

Schlichting, H.: *Boundary Layer Theory* (translated by J. Kestin) Pergamon Press, New York (1955) 146-149.

Schulman, J. H. and Leja, J.: "Control of Contact Angles at the Oil-Water-Solid Interfaces," *Trans.,* Faraday Society (June 1954) **50,** No. 374, 598.

Sherwood, T. K. and Wei, J. C.: "Ion Diffusion in Mass Transfer Between Phases," *AIChE Jour.* (1955) **4,** 1.

Sinclair, A. R.: "Heat Transfer Effects in Deep Well Fracturing," *J. Pet. Tech.* (Dec. 1971) 1484-1492; *Trans.,* AIME, **251.**

Sinclair, A. R., Terry, W. M., and Kiel, O. M.: "Polymer Emulsion Fracturing," *J. Pet. Tech.* (July 1974) 731-738.

Sinex, W. E., Schechter, R. S., and Silberberg, I. H.: "Dissolution of a Porous Matrix by a Slowly Reacting Acid," *Ind. Eng. Chem. Fund.* (1972) **11,** 205-209.

Smith, C. F., Crowe, C. W., and Nocan, T. J. III: "Secondary Deposition of Iron Compounds Following Acidizing Treatments," paper SPE 2358 presented at the SPE-AIME Eastern Regional Meeting, Charleston, W. Va., Nov. 7-8, 1968.

Smith, C. F., Crowe, C. W., and Wieland, D. R.: "Fracture Acidizing in High Temperature Limestone," paper SPE 3008 presented at the SPE-AIME 45th Annual Fall Meeting, Houston, Oct. 4-7, 1970.

Smith, C. F., Dollarhide, F. E., and Byth, N. J.: "Acid Corrosion Inhibitors — Are We Getting What We Need?," *J. Pet. Tech.* (May 1978) 737-747.

Smith, C. F. and Hendrickson, A. R.: "Hydrofluoric Acid Stimulation of Sandstone Reservoirs," *J. Pet. Tech.* (Feb. 1965) 215-222; *Trans.,* AIME, **234.**

Smith, C. F., Ross, W. M., and Hendrickson, A. R.: "Hydrofluoric Acid Stimulation — Developments for Field Application," paper SPE 1284 presented at the SPE-AIME 40th Annual Fall Meeting, Denver, Oct. 3-6, 1965.

Squier, D. P., Smith, D. D., and Dougherty, E. L.: "Calculated Temperature Behavior of Hot Water Injection Wells," *J. Pet. Tech.* (April 1962) 436-440; *Trans.,* AIME, **225.**

Standing, M. B.: *Volumetric and Phase Behavior of Oil Field Hydrocarbon Systems*, Society of Petroleum Engineers of AIME, Dallas (1977).

Staudt, J. G. and Love, W. W.: "Sustained Action Acid," *World Pet.* (May 1938) 78-80.

Sump, G. D. and Williams, B. B.: "Prediction of Wellbore Temperatures During Mud Circulation and Cementing Operations," *Trans.*, ASME (1973) **95B,** 1083-1092.

Sutton, G. D. and Lasater, R. M.: "Aspects of Acid Additive Selection in Sandstone Acidizing," paper SPE 4114 presented at the SPE-AIME 47th Annual Fall Meeting, San Antonio, Oct. 8-11, 1972.

T

Tannich, J. D.: "Liquid Removal From Hydraulically Fractured Gas Wells," *J. Pet. Tech.* (Nov. 1975) 1309-1317.

Taylor, D. B. and Plummer, R. A.: "Gas Well Stimulation Using Coiled Tubing and Acid with a Mutual Solvent," paper SPE 4115 presented at the SPE-AIME 47th Annual Fall Meeting, San Antonio, Oct. 8-11, 1972.

Templeton, C. C., Richardson, E. A., Karnes, G. T., and Lybarger, J. H.: "Self-Generating Mud Acid," *J. Pet. Tech.* (Oct. 1975) 1199-1203.

Templeton, C. C., Street, E. H., Jr., and Richardson, E. A.: "Dissolving Siliceous Materials With Self-Acidifying Liquid," U.S. Patent No. 3,828,854 (1974).

Terrill, R. M.: "Heat Transfer in Laminar Flow Between Parallel Porous Plate," *Int. J. Heat Mass Transfer* (1965) **8,** 1491-1497.

Tinsley, J. M., Williams, J. R., Tiner, L. R., and Malone, W. T.: "Vertical Fracture Height — Its Effect on Steady-State Production Increase," *J. Pet. Tech.* (May 1969) 633-638; *Trans.*, AIME, **246.**

Truse, A. S.: "Compressibility of Natural Gases," *Trans.*, AIME (1957) **210,** 355-357.

V

van Domselaar, H. R., Schols, R. S., and Visser, W.: "An Analysis of the Acidizing Process in Acid Fracturing," *Soc. Pet. Eng. J.* (Aug. 1973) 239-250.

van Poollen, H. K.: "How Acids Behave in Solution," *Oil and Gas J.* (Sept. 25, 1967) **65,** 100-102.

van Poollen, H. K. and Jorgon, J. R.: "How Conditions Affect Reaction Rate of Well-Treating Acids," *Oil and Gas J.* (Oct. 21, 1968) **66,** 84-91.

Vinograd, J. R. and McBain, J. W.: "Diffusion of Electrolytes," *Jour.*, ACS (1941) **63,** 2009.

Vitagliano, V.: "Diffusion in Aqueous Acetic Acid Solutions," *Jour.*, ACS (1955-56) **78,** 4538.

W

Wall, F. T.: *Chemical Thermodynamics*, W. H. Freeman and Co., San Francisco (1965) 372-375, 401-410.

Whan, G. A. and Rothfus, R. R.: "Characteristics of Transition Flow Between Parallel Plates," *AIChE Jour.* (June 1959) 204-208.

Wheeler, J. A.: "Analytical Calculations of Heat Transfer from Fractures," paper SPE 2494 presented at the SPE-AIME Improved Oil Recovery Conference, Tulsa, April 13-15, 1969.

Whitsitt, N. F. and Dysart, G. R.: "The Effect of Temperature on Stimulation Design," *J. Pet. Tech.* (April 1970) 493-502; *Trans.*, AIME, **249.**

Whitsitt, N. F., Harrington, L. J., and Hannah, B.: *A New Approach to Deep Well Acid Stimulation Design*, The Western Co., Fort Worth (June 10, 1970).

Williams, B. B.: "Fluid Loss From Hydraulically Induced Fractures," *J. Pet. Tech.* (July 1970) 882-888.

Williams, B. B.: "Hydrofluoric Acid Reactions with Sandstone Formations," *J. Eng. Ind.* (Feb. 1975) 252-258; *Trans.*, ASME, **97.**

Williams, B. B., Gidley, J. L., Guin, J. A., and Schechter, R. S.: "Characterization of Liquid/Solid Reactions — The Hydrochloric Acid/Calcium Carbonate Reaction," *I. and E. C. Fund.* (1970) **9,** 589-596.

Williams, B. B. and Nierode, D. E.: "Design of Acid Fracturing Treatments," *J. Pet. Tech.* (July 1972) 849-859; *Trans.*, AIME, **253.**

Williams, B. B., Nieto, G., Graham, H. L., and Leiback, R. E.: "A Staged Fracturing Treatment for Multisand Intervals," *J. Pet. Tech.* (Aug. 1973) 897-904.

Williams, B. B. and Whiteley, M. E.: "Hydrofluoric Acid Reaction with a Porous Sandstone," *Soc. Pet. Eng. J.* (Sept. 1971) 306-314; *Trans.*, AIME, **251.**

Wilson, J. R.: "Well Treatment," U.S. Patent No. 1,990,969 (Feb. 12, 1935).

Woodruff, R. A., Baker, J. R., and Jenkins, R. A., Jr.: "Corrosion Inhibition of Hydrochloric-Hydrofluoric Acid/Mutual Solvent Mixtures at Elevated Temperatures," paper SPE 5645 presented at the SPE-AIME 50th Annual Fall Technical Conference and Exhibition, Dallas, Sept. 28-Oct. 1, 1975.

Z

Zheltov, J. P. and Khristianovich, S. A.: "The Hydraulic Fracturing of an Oil-Producing Formation," *Otdel Tekh Nauk*, Izvest. Akad. Nauk USSR (May 1955) 3-41.

Author Index

(List of authors referred to in the Monograph text, references, and the bibliography.)

A

Adamson, A. W., 85, 112
Ault, J. W., 27, 28, 75, 85, 113, 115

B

BJ Hughes Inc., 67, 112
Baddour, R. F., 27, 114
Baker, J. R., 102, 117
Bancroft, D. H., 4, 112
Barenblatt, G. I., 41, 51, 112
Baron, G., 39, 51, 112
Barron, A. N., 24, 28, 47, 48, 52, 112
Beal, C., 67, 112
Beecroft, W. H., 37, 52, 114
Berry, P. M., 52, 114
Binder, G. G., 7, 9, 67, 116
Bird, R. B., 24, 28, 52, 112
Blumberg, A. A., 20, 27, 112
Bolz, R. E., 27
Bombardieri, C. C., 37, 52, 67, 90, 91, 112, 115
Boomer, D. R., 27, 28, 112
Boyer, R. F., 37, 52, 112
Brewer, L., 18, 115
Broaddus, G. C., 37, 49, 52, 67, 112, 113
Brown, R. W., 103, 112, 115
Bruce, W. A., 75, 116
Burdine, N. T., 75, 112
Byth, N. J., 102, 116

C

Cannon, G. E., 4, 112
Carpenter, N. F., 103, 112
Chamberlain, L. C., 37, 52, 112
Chang, C. Y., 24, 28, 112
Chapman, M. E., 3, 112
Chatlelain, J. C., 18, 112
Chew, J. N., 67, 112
Chittum, J. F., 90, 91, 113
Clark, J. B., 103, 112
Clason, C. E., 3, 112
Claycomb, J. R., 43, 52, 112
Connally, C. A. Jr., 67, 112
Cooke, C. E., 8, 9, 112
Couch, E. J., 52, 114
Coulter, A. W., 103, 113, 115
Crawford, P. B., 103, 115
Crowe, C. W., 37, 103, 113, 115, 116
Cunningham, J. R., 75, 85, 115

D

Daneshy, A. A., 39, 41, 51, 113
Day, F. H., 18, 113
de Groot, S. R., 27, 113
de Klerk, F., 36, 39-42, 51, 60, 67, 113
Denbigh, K. G., 18, 113
Dilgren, R. E., 90, 91, 113
Dill, W. R., 35, 37, 52, 113
Dobbs, J. B., 103, 115
Dollarhide, F. E., 102, 116
Dorsey, N. E., 67, 113
Dougherty, E. L., 52, 116
Dow Chemical Co., 2, 3, 113
Dowell, 2, 3, 67, 113, 115
Draper, H. D., 85, 113
Dullien, F. A. L., 68, 75, 113
Dunlap, P. M., 97, 103, 113

Dunlop, P., 52, 113
Dunn, L. A., 28, 113
Dysart, G. R., 44, 52, 113, 117

E

Eickmeier, J. R., 37, 43, 52, 113
Ernst, E. A., 103, 112

F

Farley, J. T., 78, 84, 113
Fast, C. R., 41, 42, 51, 52, 67, 98-100, 103, 112, 114
Fatt, I., 90, 91, 113
Finkle, P., 85, 113
Fitzgerald, P. E., 3, 113
Flood, H. L., 4, 113
Fogler, H. S., 20, 27, 28, 72, 73, 75, 85, 112, 113, 115
Foshee, W. C., 103, 113
Frasch, H., 1, 3, 113
Fredrickson, S. E., 67, 113
Fry, W. A., 3, 116

G

Gallus, J. P., 100, 103, 113
Garland, C. W., 28, 113
Gatewood, J. R., 79, 85, 113
Geertsma, J., 36, 39-42, 51, 60, 67, 113
Gidley, J. L., 28, 52, 68, 69, 75, 80, 84, 85, 90, 91, 96, 103, 113, 116, 117
Gill, W. N., 45, 52, 113
Gilliland, E. R., 22, 27, 114
Glover, M. C., 20, 27, 70, 75, 114
Goldsmith, J. R., 18, 114
Goldstein, D. J., 27, 114
Goodman, L. E., 41
Gournay, L. S., 75, 112
Graham, H. L., 103, 117
Graham, J. W., 37, 114
Grebe, J., 2
Gregory, D. P., 27, 28, 114
Grubb, W. E., 103, 113
Guin, J. A., 20, 22, 24, 27, 28, 37, 44, 45, 47, 52, 69-71, 75, 90, 112, 114, 116, 117

H

Haafkins, 39
Hall, B. E., 84, 85, 96, 97, 102, 113, 114
Halliburton Oil Well Cementing Co., 2-4, 113
Ham, W. E., 18, 114
Handbook of Chemistry and Physics, 18, 114
Handbook of Tables for Applied Engineering Science, 114
Hannah, B., 52, 117
Hanson, H. R., 113
Harned, H. S., 22, 27, 114
Harrington, L. J., 52, 117
Harris, F. N., 14, 18, 114
Harrison, N. W., 52, 99, 103, 114
Hathorn, D. H., 3, 114
Hazebrock, P., 9, 116
Hendrickson, A. R., 4, 28, 48, 52, 84, 103, 112, 114, 115, 116
Herrington, C. G., 4, 114
Hidy, M. E., 3, 114
Hidy, R. W., 3, 114
Hildebrand, J. H., 85, 113
Hill, W. L., 52, 114
Howard, G. C., 41, 42, 51, 52, 67, 98-100, 103, 112, 114
Hurst, R. E., 103, 113

I

International Critical Tables, 18, 67, 114

J

Jazraui, W., 90, 114
Jenkins, R. A., Jr., 102, 117
Jorgon, J. R., 52, 117
Joyner, H. D., 9, 114
Juch, A. H., 67, 103, 115

K

Karnes, G. T., 85, 115, 117
Keeney, B. R., 103, 114
Keller, H. H., 43, 52, 114
Kern, L. R., 51, 116
Kerver, J. K., 37, 114
Khristianovich, S. A., 41, 51, 114, 117
Kiel, O. M., 39, 51, 67, 114, 116
Klock, G. O., 75, 114
Knox, J. A., 37, 49, 52, 67, 103, 112, 114
Knox, T. A., 37, 114
Krech, W. W., 51, 116
Kruk, K. F., 34-37, 50, 52, 63, 88, 91, 103, 115
Kucera, C. H., 67, 114

L

Labrid, J. C., 17, 18, 76, 80, 84, 114
Lagrone, K. W., 67, 103, 114
Laidler, K. J., 19, 27, 114
Lasater, R. M., 85, 97, 103, 113, 114, 117
Leiback, R. E., 103, 117
Leja, J., 85, 116
Lenhard, P. J., 18, 115
Leutwyler, K., 43, 52, 115
Levich, V. G., 26, 28, 115
Lewis, G. N., 18, 115
Lightfoot, E. N., 28, 52, 112
Loper, R. G., 103, 112
Love, W. W., 52, 117
Lovingfoss, W. J., 9, 114
Lund, K., 20, 27, 28, 75, 85, 113, 115
Lybarger, J. H., 85, 115, 117

M

Malone, W. T., 9, 117
Martin, F. G., 90, 103, 114
Martin, T. H., 91, 112
Mayhill, T. D., 85, 113
Mazur, P., 27, 113
McBain, J. W., 22, 27, 117
McCann, B. E., 3, 115
McCune, C. C., 27, 28, 71-73, 75, 79, 85, 112, 113, 115
McDougall, L. A., 102, 115
McGinty, J. E., 103, 115
McGuire, W. J., 7, 67, 115
McLeod, H. O., 103, 115
McPherson, A. M., 3
Miller, B. M., 84, 113
Miller, D. G., 22, 27, 115
Montgomery, P., 3
Moore, E. W., 103, 115
Moore, W. W., 4, 115
Morgan, F. A., 37, 114
Morian, S. C., 4, 115
Morse, R. A., 9, 115
Moss, J. T., 43, 52, 115
Mowrey, S. L., 20, 27, 115
Murphy, W. B., 67, 103, 115
Muskat, M., 6, 9, 115

N

Nathan, C. C., 103, 112
Neill, G. H., 103, 112, 115
Neuman, F. M., 90, 91, 113
Newcombe, R. B., 3, 115
Nierode, D. E., 20, 22, 24, 25, 27, 34-37, 44-52, 60, 61, 63, 67, 73-75, 87, 88, 91, 103, 115, 117
Nieto, G., 103, 117
Nocan, T. J., III, 103, 116
Nordgren, R. P., 51, 115
Notter, R. H., 23, 28, 115
Novotny, E. J., 50-52, 100, 103, 115, 116

O

Oakes, B. D., 102, 116
Owens, B., 22, 27, 114

P

Perkins, T. K., 51, 116
Perrine, R. L., 90, 91, 116
Perry, J. H., 18, 116
Pickering, S. V., 85, 116
Pitzer, K. S., 18, 115, 116
Pitzer, P. W., 4, 116
Plummer, R. A., 97, 103, 117
Pollock, R. W., 37, 52, 114
Prats, M., 7, 9, 116
Pruitt, G. T., 103, 115
Purcell, W. R., 75, 116
Putnam, S. W., 3, 116
Pye, D. S., 100, 103, 113

R

Ramey, H. J., Jr., 37, 43, 52, 113, 116
Randall, M., 18, 115
Rasmussen, J. W., 67, 103, 114
Raymond, L. R., 7, 9, 43, 52, 65, 67, 116
Reichertz, P. P., 75, 112
Resibois, P. M. V., 27, 116
Richardson, E. A., 85, 117
Riddiford, A. C., 27, 28, 114
Rike, J. L., 105, 106, 116
Roberts, H., 3, 4, 116
Roberts, L. D., 22, 24, 27, 28, 37, 44, 45, 47, 52, 85, 112, 113, 116
Robinson, R. A., 18, 116

Rose, W. D., 75, 116
Rosene, R. B., 28, 52, 114
Ross, W. M., 84, 103, 115, 116
Rothfus, R. R., 52, 117
Rowan, G., 90, 116
Russell, D. G., 9, 116
Ryan, J. C., 85, 113

S

Sarmiento, R., 67, 116
Schechter, R. S., 18, 28, 52, 68, 69, 71, 75, 90, 112, 114, 116, 117
Scheidegger, A. E., 68, 75, 116
Scheuerman, R. F., 85, 115
Schlichting, H., 24, 28, 116
Schoettle, W., 84, 113
Schols, R. S., 52, 117
Schulman, J. H., 85, 116
Schweppe, J. L., 43, 52, 112
Sherwood, T. K., 22, 27, 116
Sikora, V. J., 7, 9, 67, 115
Silberberg, I. H., 18, 75, 112, 114, 116
Simmons, C. M., 103, 115
Sinclair, A. R., 33, 37, 39, 43, 51, 52, 67, 116
Sinex, W. E., 70, 75, 116
Sleicher, C. A., 23, 28, 115
Smith, C. F., 4, 35, 37, 76, 84, 93, 95, 101-103, 115, 116
Smith, D. D., 52, 116
Smithey, C. M., 103, 113
Squier, D. P., 43, 52, 116
Standing, M. B., 67, 117
Staudt, J. G., 3, 52, 112, 117
Stavrinou, S. C., 20, 27, 112
Stewart, W. E., 28, 52, 112
Stockmayer, W. H., 28, 113
Stokes, R. H., 18, 28, 113, 116
Street, E. H., Jr., 85, 117
Strickler, W. R., 9, 115
Sump, G. D., 43, 52, 117
Sutton, G. D., 85, 97, 103, 117

T

Tannich, J. D., 8, 9, 102, 103, 117
Taylor, D. B., 96, 103, 117
Templeton, C. C., 83-85, 117
Terrill, R. M., 45, 52, 117

Terry, W. M., 67, 116
Texas Petroleum Research Committee, 27, 28, 75, 114, 115
The Western Co., 67, 116, 117
Thomas, W. A., 2
Tiner, L. R., 9, 117
Tinsley, J. M., 7, 9, 117
Tomer, F. S., 90, 91, 113
Tong, S., 28, 113
Truitt, N. E., 9, 116
Truse, A. S., 67, 117
Tuve, G. L., 27

V

Vandergrift, J. G., 2
van Domselaar, H. R., 44, 45, 52, 117
van Dyke, J. W., 1
van Poollen, H. K., 52, 117
Vinograd, J. R., 22, 27, 117
Visser, W., 52, 117
Vitagliano, V., 28, 117
Von Gonten, W. D., 9, 115

W

Wahl, H. A., 52, 114
Wall, F. T., 27, 117
Weaver, R. H., 67, 114
Wei, J. C., 22, 27, 116
Wescott, B., 2
West, C. K., 4, 116
Whan, G. A., 52, 117
Wheeler, J. A., 43, 44, 52, 117
White, P. D., 43, 52, 115
Whiteley, M. E., 71, 73, 75, 78, 79, 85, 116
Whitsitt, N. F., 44, 52, 113, 117
Wieland, D. R., 28, 37, 48, 52, 112, 114, 116
Williams, B. B., 20, 22, 24, 25, 27, 28, 37, 42-52, 60, 61, 67, 71-73, 75, 78, 79, 85, 87, 91, 103, 115, 116
Williams, J. R., 9, 117
Wilson, J. R., 2-4, 117
Woodruff, R. A., 102, 117

Z

Zheltov, J. P., 41, 51, 114, 117

Subject Index

A

Absolute rate theory, 19
Acetic acid:
  Applications in well completion, stimulation and reconditioning, 18
  As iron sequestering agent, 102
  Diffusion, 28
  Diffusion coefficient, 23
  Dissociation constant, 14, 15
  Formed from acetic anhydride and water, 90
  In carbonate acidization, 10, 11, 13
  In low permeability formation, 88
  Preventing precipitation of ferric hydroxide, 101
  Retarded acid system, 35
Acetic anhydride, 90
Acetic-hydrochloric acid, 11
Acetylenic alcohols, as inhibitor, 92
Acid additives, 92-103
Acid balance equation, 71-73
Acid concentration profile, 34, 44, 73
Acid concentration ratio, 46, 73, 74
Acid contact time, 35
Acid corrosion rate, effect of metal type on, 93, 94
Acid etch pattern, 36
Acid-external emulsion fluid, 31
Acid flow behavior:
  Effect of rate on Berea core response to HF-HCl, 77
  In fracture, 30
Acid fracturing:
  Additives, 88
  Analysis of acidizing process in, 52
  As acidizing technique, 5, 7-9
  Development of stable polymers, 12
  Diverting agents, 100
  Economic analysis, 106
  Friction reducers, 97
  Fundamentals, 29-38
  Predicting acid penetration distance, 74
  Reducing corrosion rate, 95
  Treatment design, 53-67
  Treatment models, 38-52
  Using cellulose-based fluids, 59
  Using organic acid anhydrides, 90
Acid mixtures, 10, 11
Acid penetration distance:
  Along a circulating tube, prediction of, 26
  Along a fracture, 26, 30-35, 38, 44-49
  Calculation of, 60-63, 66
  Into sandstone, prediction of, 71-74
  Limited without fluid-loss additive, 88
  Maximum, no fluid loss, 87
  Method for predicting, 27, 52, 79
  Predicting for the fracturing fluid and acid of interest, 88
Acid polymer emulsion, 51
Acid reaction:
  Along the fracture, models for, 44
  Distance, predicting, 44, 60-62
  In limestone formations, 52, 73-75, 91
  Kinetic models for, 20, 27, 50
  Prediction of radius, 78, 79
  Rate, 12, 24, 88
  Rate retarders, 96
  Rate tests, 78
  With carbonates, 12, 73
Acid response curve, 76, 77, 79
Acid-sensitive viscous gel, 5
Acid solvents, for oil wells, 52
Acid sticks, 11

Acid stimulation design, for deep wells, 52
Acid-swellable polymers, diverting agents, 99
Acid treatments:
  Beneficial to North Louisiana gas wells, 4
  Description of, 5
  First, 1
  Of lime wells, 4
  Oil wells, 3
Acid-volume-to-coupon-area ratio, 94, 95
Acid washing, 5
Acidization reactions, for kinetic models, 20
Acidizing:
  Argillaceous sandstone, 18
  Carbonates, 19, 86-91
  Economics, 104-106
  Formation of companies, 2
  Frasch patent, 1
  Matrix, 68-91
  Methods, 5-9
  Models for, 68-75
  Modern era of, 2
  Patents on, 1
  Sandstones, 76-85
Activity coefficient:
  of hydrochloric acid, 14
  of mixtures, 21
Afterflush, 1, 76, 80-83, 86, 96
Agitation, effect on acid corrosion inhibitor effectiveness, 93
Albite, hydrochloric acid and hydrofluoric mixtures, 20
Alcohols:
  As cleanup additive, 102
  Low molecular weight, 81
  Reaction product, 90
  Use in gas well stimulation, 103
Alkyl aryl sulfonate, surfactant, 96
Alkyl phosphates, anionic surfactant, 96
Allyl chloride, 90
Ammonium chloride spacer, 84
Ammonium fluoride, 83
Antisludge agents, 96
Applied engineering stimulation, 67
Arrhenius equation, 21
Arsenic compounds, as inhibitors, 92, 95
Asphaltene sludges, 103
Autoclave, 93

B

Ball sealers, 66, 100, 101, 103
Banada Field, 17
Barbers Field, 17
Benzoic acid, 100
Bottom-hole injection temperature, 43
Boundary layer theory, 28
Breakdown pressure, 55, 67
Brine viscosity, 57
Byron-Jackson, Inc., 2

C

Calcitic rocks, 12
Calcium acetate:
  Product of reaction with acetic acid, 11
  Solubility in water, 16
Calcium carbonate:
  And acetic acid system, 15
  Inert solid for matrix diversion, 99
  Reaction with HCl, 12, 13, 19, 20, 28, 52, 75
  Reaction with HF, 76
  Solid fluid-loss additive, 98

Calcium chloride:
  Molecular weight, 13
  Soluble product in Frasch acidizing patent, 1
  Soluble reaction product of HCl, 10, 12
Calcium formate, solubility in water, 16
Calcium sulfate, insoluble product of Van Dyke process, 1
California, heavy asphaltic oils, 96
Canada, heavy asphaltic oils, 96
Carbon dioxide:
  Acid reaction kinetics, 21
  Cleanup of well after stimulation, 59
  Created by HCl reaction, 35
  Molecular weight, 13
  Results of well stimulation, 103
  Soluble product of acid treatment of carboneous formations, 10, 12
  Soluble product of Frasch patent, 1
  Solubility in water, 16
Carbonate acidization, acids used in, 13
Carbonate rocks:
  Classification of, 18
  Formation of, 12
  Measured rock embedment strength of, 50
Cash flow analysis, 104, 105
Cellulose additive, 97-99
Cellulose-based fluids, 59
Chemical activity, definition, of, 19
Chemical Engineers Handbook, 18, 116
Chemical Process Co., 2
Chemically retarded acids, 12, 35, 52, 66, 88
Chloroacetic acid, 10, 11, 23, 28
  Diffusion coefficient, 23
  Diffusion in water, 28
  Powered organic acid, 10, 11
Citric acid, sequestering agent, 101, 102
Clastic grains, 12
Clean-out acids, 96
Cleanup additives, 102
Cleanup time, 8
Closure stress, 35, 50, 51, 63
Compatibility tests, 81
Complexing agents, 100-102
Compressibility:
  Brine, 57
  Formation fluid, 56
  Gas, 58
  Isothermal coefficient of, 56
  Natural gases, 58, 67
  Oil, 56, 57
  Pseudo-reduced, 58
  Reduced, 57
  Water, 57
Compressive earth stresses, 7
Compressive strength, of formation, 78
Computer model, for fracture geometry, 39
Conductive-fracture length, 5
Conductivity ratio, 7, 29
Connate water, 10
Contact angles, 85
Convection transfer, 21-23
Convective mixing, 74
Conversion factors, for quantities having dimensions of:
  Concentration times diffusivity, 108
  Density times diffusivity, 108
  Energy, 108
  Force, 107
  Fugacity, 107
  Heat-transfer coefficients, 109
  Mass-transfer coefficients, 109

Molecular diffusivity, 108
Momentum diffusivity, 108
Pressure, 107
Thermal conductivity, 108
Thermal diffusivity, 108
Torque, 108
Viscosity, 108
Work, 108
Corrosion inhibitor:
    Absorption on surfaces, 80
    For high temperature acid, 102
    In preflush, 76
    Selection, 59, 89
    To protect wellbore tubulars, 86, 92-95
Corrosive liquid environment, 28
Corrosivity, wellbore tubular goods, 11
Crosley Farm, application of acidizing
    process, 1

D

Damage induced by acid, 76
Darcy's law, 50
De-emulsifiers, 86
Delaware-Val Verde basins, 52
Demobilization costs, 104
Demulsifying agent, 95, 96
Diffusion:
    Coefficients, 21-23, 28, 35, 45, 47
    Convection, 44, 52
    Counter, 27
    Experimental coefficients, 22
    Ion, 27
    Ionic coefficient, 46
    Molecular coefficient, 74, 75
    Of electrolytes, 27
    Of strong acids, 21
    Of weak acids, 27
Diffusivity:
    Eddy, 23, 28
    Effective, 22
    Equation, 43
Dimensionless conversion rate, 25, 26
Discounted cash flow techniques, 106
Discounted rate-of-return, 104
Dissociation constant, 14, 15
Dissolving power of:
    Acetic acid, 13, 14
    Acid, 10, 49, 59, 63
    Formic acid, 13
    Hydrochloric acid, 13
    Hydrofluoric acid, 16, 78
    Oil external emulsions, 34
Distribution function, 68
Diverting agent, 66, 99, 100, 103
Drainage radius, definition, 6
Drilling mud:
    Plugging formation pores, 80
    Removal from formation by acid, 3-6
    X-ray analysis, 17

E

East Bay fields, 84
Economic evaluation, 53, 104, 106
Economic limit, 104
Effective mass-transfer coefficient, 23, 74
Electrolyte solutions, 18, 27
Emulsification, theory of, 85
Emulsified acids:
    Designing treatments using, 66
    Retarded acid system, 10, 12
    Use in high-permeability or naturally
        fractured formation, 89
    Use in matrix treatments, 88
    Viscous acid system, 34
Emulsions:
    Between reservoir oil and reacted acid, 80
    Eliminate formation, 92

Use of acids to breakdown, 5
    Viscosity, 66
    Viscous, 59
Engineering Data Book, 67, 113
Equilibrium:
    Calculation of, 15
    Chemical, 14, 18
    Concentration, 11
    Constant, 14, 20
    Cracks, 51
    Dissociation, 14
    Fracture geometry, 38-41
    In-acid-carbonate reactions, 14
    In-acid-sandstone reactions, 17, 18, 76
    Reaction, 14, 20, 21
    Thermodynamics, 10
Equipment mobilization, 104
Ethoxylated alkylphenol, surfactant, 96
Ethylene diamine tetracetic acid, 101, 102
Ethylene glycol monobutyl ether (EGMBE),
    81-83, 96, 97
Ethylene oxide substitution, 96
Exposure time, 93, 94

F

Ferric hydroxide, 101
Ferric iron, 100
Ferrous iron, 100, 101
Fick's law, 22
Finite-difference model, 43
Fluid injection temperature, 56, 60
Fluid loss:
    Control, 58, 97, 98
    From fractures, 41-45, 48
    Limit, 87
    Reynolds number, 62
    Test, 31, 87, 88, 98
Fluid-loss additive:
    Changes in fluid-loss characteristics, 32,
        42
    Concentration, 61
    Effect of concentration on wormhole
        growth rate, 88
    Evaluation of, 37, 52
    For acid, 59, 65, 66, 89, 91, 97, 98,
        103
    Fracturing, 99
    Maximizing acid penetration distance, 31
    Mechanism by which acid bypasses, 30
    Reducing fluid loss from a wormhole, 87
    Water-based pad fluids, 59
Fluid-loss coefficient, 40-42, 61, 66, 97
Fluid temperature, in the fracture, 42-44
Fluid spurt, 42
Fluoaluminates, 18
Fluoride ion concentration, 17
Fluorine ion concentration, 17
Fluosilicate reaction products, 76
Fluosilicic acid, 17
Formation damage, 6, 8, 90
Formation elastic properties, 38
Formation fluid density, 58
Formation fluid viscosity, 56
Formation fracturing gradient, 55
Formation matrix, 53
Formic acids:
    Calcium carbonate system, 15
    Constants for determining acid dissociation
        constants, 14
    Dissociation coefficients, 23
    Dissolving power, 13
    Hydrolysis of methyl formate to form, 83,
        84
    Organic acids, 10, 11
    Reaction with ammonium fluoride to yield
        HF, 83
    Retarded acid system, 35

Used in matrix treatments, 88
    Value for acid dissociation constant, 15
Formic-hydrochloric acid, 11
Formic-hydrofluoric acid, 11
Forward rate constant, 19
Frac Guide Data Book, 113
Fracbook Design/Data Manual, 113
Fracture conductivity:
    Assuring adequacy, 34
    Controlled by treatment design and
        formation strength, 29
    Evaluation of, 37, 52, 91, 103
    Influenced by volume of rock dissolved,
        rock strength and closure stress, 35, 36
    Maximize, 66
    Predicting, 49-53, 62-65
Fracture design chart, 39
Fracture-fluid invasion equation, 41
Fracture geometry:
    Acid fracturing treatment, 29-31
    Dynamic, 38-40
    Estimate of fluid temperature needed to
        predict, 32
    Equilibrium, 38, 40-42
    For example limestone formation, 61
    In Barron, et al. model, 48
    Model, 39, 41
    Predict, 53, 55, 60
Fracture gradient, 53, 55, 82, 89
Fracture orientation, influence on mixing,
    45
Fracture propagation pressure, 29, 30
Fracture proppants, conductivity of, 9
Friction-pressure data, 59, 60, 96
Friction reducers, 96, 97

G

Gas reservoirs, acid fracturing, 7, 8
Gas viscosity, as function of specific
    gravity, temperature, and pressure, 57
Gas wells, performance of, 9
Gaseous nitrogen, 102
Gelled acids, 10, 34, 54
Gelled waters, 33, 59
Gelling agents, 12, 34, 59
Geochemistry, researches in, 18
Geologic risk factor, 105
Gluconic acid, sequestering agent, 101, 102
Glycol ethers:
    Low molecular weight, 81
    Mutual solvent, 96
Golden Meadows Field, 17
Gulf Coast:
    Area, 3
    Sandstone formations, 17
Gulf Oil Co., 2
Guar:
    Acid reactions at high temperatures, 99
    As fluid-loss additive, 98
    As friction-reducer additive, 97
    In gelled acids, 34
    In gelled water, 59
Gum karaya, 34, 97-99
Gypsy Oil Co., 2

H

Heat-transfer coefficient, 43
Heterogeneous reaction, 19
HF-HCl reaction, effect on core mechanical
    properties, 78
Hot water in wells, calculated temperature
    behavior, 52
Hybrid acid mixture, 15, 16
Hydraulic fracture design, effect of
    formation temperature, 52
Hydraulic fracturing:
    Monograph, 38, 52, 67, 103

Propagation of, 51
Treatments, 8
Well stimulation, 1
Hydrochloric acid:
  Acid concentration, 59
  Acid dissociation constant, 15
  Activity coefficient, 14
  Complexing agents in, 101, 102
  Diffusion, 21, 22
  Dissolution of calcite in, 27
  Dissolving power, 13
  Effect of acid volume/coupon area on
    corrosion rate, 95
  Example calculation of acid fracturing
    treatment, 60-63
  Frasch acidizing, patent, 1
  Inhibitors for, 2, 10, 92-95, 102, 103
  In-situ acid formation, 90
  Internal phase of emulsified acid, 12
  Mixing coefficients, 47, 74
  Mixture with acetic acid, reaction times
    of solutions, 52
  Mixture with HF, 6, 16, 17, 20, 28,
    76-85, 94, 102
  Reaction with limestone, 12, 13, 19, 20,
    28, 45, 52, 75, 86
  Specific gravity of aqueous solutions, 13
  Spending time data for reaction with
    limestone, 48, 49
  Treatment of carboneous formations, 10,
    11
  Tubular friction pressure, 96
  Well treatment, 2
Hydrofluoric acid:
  Concentration profile, 71
  Corrosion in mud acid, 94
  Dissolving power, 16
  For treating sandstones, 2, 4
  Mixture with HCl, 3, 6, 10, 11, 16-18,
    20, 28, 102
  Mixture with HCl, for treating sandstones,
    76-85, 96
  Reaction with sandstone, 75
  Reaction with vitreous silica, 20, 27, 69
Hydrogen chloride:
  Gas, 10
  Ionization, 14
Hydrogen embrittlement, 95
Hydrogen fluoride, 3, 83
Hydrogen sulfide, 95
Hydrofluosilicic acid, 3
Hydrostatic gradient, 81-83, 89

I

Ideal fracture conductivity, 36, 50, 64
Ideal fracturing fluid, 58
Imperial Oil Ltd., 86
Inhibitor:
  Acetylenic alcohols, 92
  Acid corrosion, evaluation of, 93
  Arsenic acid, 2
  Arsenic compounds, 92
  Extender, 95
  Nitrogen heterocyclics, 92
  Regular acid, 10
  Rodine No. 2, 2
  Rosin amine derivatives, 92
  Substituted thioureas, 92
  Suggestions for selection, 95
Injectivity ratio, 99
Injectivity falloff test, 54
Integral fluid-loss equations, 42
Ion transport coefficients, 27
In-situ acid formation, 90
Iron chelating agents, 86
Irreducible water, 80
Isopropyl alcohol, 90

K

Kansas, limestone reservoir rocks, 3
Kinetic models, for acidization reactions, 20
Kinetic parameters, for acid reactions, 21
King Royalty Co., 3

L

Lactic acid, 101, 102
Lake St. John, 17
Laminar flow:
  Between parallel porous plates, 52
  Circular tube reactor, 25
  Models for heterogeneous reactions, 23-27
  Nierode and Williams model, 45
  Parallel plate reactors, 26
  Rate of mass transfer, 22
Leakoff:
  Rate, 44, 45
  Reynolds number, 61
Limited entry technique, 67, 100, 103
Loma Novia Field, 17
Louisiana, North, acid results, 3
Low surface tension acid, 96

M

Magnesium acetates, 11
Magnesium chloride, 13
Manvel Field, 17
Mass-transfer coefficient, 44, 45
Mass-transfer model, 21
Matrix acidization:
  Description of model for, 68, 69
  Gelling agents, 12
  Of carbonates, 88, 89
  Theory of, 27, 73
  With highly reactive acids, 75
Matrix acidizing:
  Diverting agents, 99, 100, 103
  Of carbonates, 86-91
  Of sandstones, 76-85
Matrix composition, effect on core response
  to HF-HCl, 77
Matrix fracturing, 5-7
Mechanical risk factor, 105, 106
Mechanism of acid corrosion inhibition, 92
Mellon Institute, 2
Metal area/acid volume ratio, 93
Methyl formate, 83, 84
Michigan:
  Exploration program, 2
  Oil Maker's Co. formed, 2
Milk of lime, 1
Mineral acids, 10, 11, 13
Mixing coefficients:
  Effective, 44, 46, 47, 62, 74
  For oil-wetting surfactants, 35
Models, acid fracturing treatment:
  Barron and Wieland, 47, 48
  Eickmeier and Ramey, 43
  Fogler-McCune, 72, 73, 75
  For matrix acidizing, 68-75
  Geertsma and de Klerk, 39-42
  McCune, et al., 71-73, 79
  Nierode-Williams, 45-50, 61, 63, 73
Models, acid fracturing treatment:
  Roberts and Guin, 44, 45, 47
  Surface kinetic, 69
  Wheeler, 43, 44
  Whitsitt and Dysart, 44
  Williams and Whiteley, 71, 73, 79
Molecular diffusion coefficient, 74, 75
Monocarboxylic acids, 28
Monte Carlo methods, 71
Mud Acid:
  Concentration profile, 71

Depth of permeability increase, 79
Dissolution of selected minerals, 27
First commercial use, 3, 4
Inadequate volume, 83
Self-generating treatments, 84, 85
Spent, 80
Theory and application to oil and gas wells,
  18
Muriatic acid, 1
Mutual solvents:
  Acid method, 84
  As additive, 86, 93, 95
  Corrosion inhibition, 102
  EGMBE, 81-83, 96, 97
  Gas well stimulation, 103

N

Naphthalene, 100, 101
Net profit-to-investment ratio, 104
Nitrilo triacetic acid, 101, 102
Nitrogen, in well stimulation, 103
Nitrogen heterocyclics, 92
Nitroglycerin, 2
Nitro-shooting, 1
Nonemulsifying acid, 96
Normal propyl chloride, 90
Numerical models, 43

O

Ohio Oil Co., 1
Oil-based pad fluids, 59
Oil City Derrick, 1, 112
Oil-external emulsion, 34
Oil Maker's Co., 2
Oil-soluble polymers, 99
Oil-soluble resins, 99
Oil-soluble waxes, as fracturing diverting
  agent, 100
Oil viscosity, gas-saturated, 56
Oklahoma:
  Hydrochloric acid treatment, 2
  Williams Brothers Treating Corp. formed,
    2
Organic acids:
  Carbonate system, 18
  Dilute, 10, 11
  In-situ acid formation, 90
  Reaction equilibrium, 14-16
  Solid, used for matrix diversion, 99, 100
  Used as complexing agent, 101
  Used in matrix treatment, 88
Organic amines, 96
Organic polymers, 97, 98
Organic resin:
  Fluid-loss additive, 98
  Fracturing diverting agent, 100
  Matrix diversion, 99
Organic sulfonate, fluid-loss additive, 98
Orthosilicic acid, 17, 77
Overburden gradient, 89

P

Pad fluid:
  Dimensions of a fracture created by, 60,
    61, 63
  Effect on fracture geometry, 29-31
  Fluid-loss test results when acid follows,
    98
  In acid fracturing, 7, 8, 53, 97
  Types, 58, 59
  Viscous, 36, 66
Paraformaldehyde flakes, 100
Parallel plate reactor, 23-27
Payout period, 104-106

Permeability:
Absolute, 80
Changes, in dissolving porous media, 75
Combined effect of overburden stress and acid throughput on, 78
Damaged, 6, 8
Formation, 7, 54, 55, 87, 89
Improvement, 70-72, 76, 79, 82
Matrix, 81
Matrix acidizing diverting agents, 99
Mechanism of acid attack, 86
Property required for acid treatment design, 53, 60
Reduction, 5, 18, 77, 100
Relation to pore size distribution, 69
Relative, 80
Permeability ratio, 6, 68, 69-71, 78
pH reduction, 5
Pitting tendency, 95
Poisson's ratio, 38, 53, 55, 56, 60
Polyacrylamide, 34, 59, 96-99
Polyisobutylene, friction reducer additive, 97
Polymer emulsion fracturing, 67
Polyoxyethylated alkylphenols, 96
Polyvinyl alcohol, fluid-loss additive, 98, 99
Pore size distribution, 68-71, 75, 90
Pore volume, 6, 77
Porosity:
Formation, 5, 6, 55
Geological factors influencing, 67
Improvement, 70
Property required for acid treatment design, 53, 60
Reaction rate to account for changes, 71-73
Relation to pore size distribution, 69
Porosity ratio, 69-71
Potassium fluosilicates, 10, 76
Potassium iodide, 95
Powered acids, 10, 11
Precipitated chert, 12
Preflush, 5, 76, 80, 82-84, 96
Pressure buildup test, 54
Pressure gradient:
Effect on core response to HF-HCl, 77
Imposed by HCl in wellbore, 82
Pressurized fracture length, 41
Product technical guide, 67
Productivity:
After sandstone acidizing, 79-81
Effect of wettability change on, 80
Improvement, theoretical from acidization, 5-9
Increase from matric acid treatment, 88
Productivity ratio, 6
Profit-to-investment ratio, 105, 106
Propagation pressure, 38, 42, 55
Propionic acid, 90
Pseudocritical properties, 57, 58
Pure Oil Co., 2

Q

Quaternary amines, 96

R

Radiation survey, 54
Radius of reaction, prediction of, 87, 88
Rate coefficients, 21
Rate of acid transfer, 19
Rate of return, 106
Ratio of:
Acid concentration, 73, 74
Acid volume to coupon area, 94
Conductivity of fracture to formation permeability, 7, 29
Damaged-zone permeability to

undamaged-formation permeability, 8
Fracture conductivity to formation permeability, 7, 65
Fracture length to drainage radius, 7
Fracture volume to injected volume, 33
Fracture width measured at wellbore to fracture length, 29
Injectivity, 99
Metal area to acid volume, 93
Net profit-to-investment, 104
Permeability of porous medium during acidization to original permeability, 6, 68, 69, 78
Porosity of porous medium during acidization to original permeability, 70
Stimulation, 6-8, 10, 51, 62-66
Surface rock area to acid volume, 48
Reaction model, 19, 20
Reaction rate:
Average, 69
Chemical factor in selecting acid, 10
Coefficient, 71, 72
Constant, 73
Limit, 73, 90
Of component in heterogeneous reaction, 19
Of well treating acids, 52
Surface, 45, 47
Reaction time, 48, 49
Reactors, 23-26
Regular acid, 10
Relative permeability curves, 80
Reservoir fluid properties, 53, 60
Reservoir risk factor, 105
Respirators for personnel, 3
Retarded acid systems:
Application of matrix acidizing model, 69, 70
Emulsified, 34
Evaluation of, 91
Mixture of organic acids and HCl, 35
Organic acids as, 90
Surfactants as, 96
Used to stimulate carbonate formations, 10, 12
Retarded reaction rate, 11
Rock embedment strength, 35, 36, 50, 63
Rock salt, 99, 100
Rocky Mountain states, heavy asphaltic oils, 96
Rosin amine derivatives, 92
Rotating disk reactor, 23, 26, 27

S

Sandstone acidizing treatment, 76, 79, 81, 83, 84
Scale removal treatment, 2
Segno Field, 17
Sheridan Field, 17
Silica flour:
Fluid-loss additive, 31, 98
For matrix diversion, 99
Siliceous fossils, 12
Silicic acid, 3
Silicon dioxide, 17
Silicon tetrafluoride, 3, 17
Sodium fluoride, 3
Sodium fluosilicates, 10, 76
Solubilization process, 17
Sonic travel time, 53, 55, 56
South Pass Block 24 Field, 84
South Pass Block 27 Field, 84
Specific gravity, of aqueous hydrochloric acid solutions, 13
Spending time data, 48
Spent acid, 59, 82, 87, 102
Spurt volume, 40, 42, 61, 98

Squeeze cementing, 5
Stabilization of emulsions by solids, 80
Standard Oil Co., 1-4
Static reaction rate test, 23, 48, 49
Steam injection wells, casing temperature studies, 52
Stimulation ratio:
Attained from damage removal alone, 66
Comparison of predicted with observed, 51
Expected, 10
Increase in production from vertical fractures, 7
Matrix acidizing of a carbonate, 88
Maximum attainable in an undamaged well with a matrix acid treatment, 6, 70
Prediction of, 62-65
Resulting from an acid fracturing treatment, 29, 97
Stoichiometric coefficients, 13, 73
Stoichiometry:
acid, 10
of acid-carbonate reactions, 12-14
of acid-sandstone reactions, 16, 17
Substituted thioureas, 92
Sulfamic acids, 10, 11
Sulfide stress cracking, 103
Sulfonates, 96
Sulfur dioxide, 90
Sulfuric acid, patent, 1
Surface active materials, 81, 95, 102
Surface kinetics:
Effect in fracture acidizing, 37
Effective parameters, 33
In acid fracturing models, 44, 47
Model, 69
Rate, 71
Surface reaction kinetics, 19-21, 27, 52
Surfactant:
Cleanup additive, 102
Oil-wetting, 12
Used in acid treating, 95, 96
Swellable polymers, 103
Synthetic water-soluble polymers, 59

T

Taurates, 96
Temperature depth profiles, 33
Temperature logs, 54
Temperature profile:
Along fracture, 43, 44
In water injection wells, 52
Tensile strength, of rock, 7
Tertiary butyl chloride, 90
Texas, Chemical Process Co. formed, 2
Thiophenols, 92
Tortuosity, 68
Toxicity data, 92
Transient pressure behavior, 9
Tubular reactor, 23, 25-27
Turbulent flow, 22, 23, 45

U

Unbounded fracture, 38

V

Vertical fracture height, 9, 53, 54
Villa Platte, 17
Viscous acid systems, 34
Viscous fingering, 36
Vitreous silica, 20, 27

W

Water-based pad fluids, 59
Water-soluble polymer, 99

124

Water solubility, 96
Wax-polymer agents, 100, 101
Weak acids, definition, 14
Well productivity:
  Comparison of actual with ideal, 8
  Effect of a damage zone adjacent to a
    fracture, 8
  Effect of vertical fractures on, 9, 67
Wellbore damage, design of acid fracturing
  treatment to remove, 66
Wellbore heat transmission, 52
Wharton Field, 17
Wildcat Field, 17
Williams Brothers Treating Corp., 2
Wormholes:
  Convective transfer, 21
  Definition, 86
  Effect of fluid-loss additive concentration
    on growth rate, 88
  Formation predicted, 71
  Formed in face of a limestone fracture
    wall, 30, 31
  Maximum length, 87
  Model of growth in carbonate acidization,
    73-75
  Pattern for matrix acid treatment in
    carbonate, 7
  Spacing, 66
  Tubular reactor, 23
  Zero fluid loss from, 25
Wyoming bentonite, 84

X

X-ray analyses, of typical Gulf Coast
  sandstone formation, 17

Y

Yield strength, specification, 93
Young's modulus, 30, 38, 40, 41, 55, 56,
  60

Z

Zwolle wells, 4